T0326477

Critical Ecofeminism

Ecocritical Theory and Practice

Series Editor: Douglas A. Vakoch, California Institute
of Integral Studies, USA

Advisory Board

Bruce Allen, Seisen University, Japan; Hannes Bergthaller, National Chung-Hsing University, Taiwan; Zélia Bora, Federal University of Paraíba, Brazil; Izabel Brandão, Federal University of Alagoas, Brazil; Byron Caminero-Santangelo, University of Kansas, USA; Simão Farias Almeida, Federal University of Roraima, Brazil; George Handley, Brigham Young University, USA; Isabel Hoving, Leiden University, The Netherlands; Idom Thomas Inyabri, University of Calabar, Nigeria; Serenella Iovino, University of Turin, Italy; Daniela Kato, Kyoto Institute of Technology, Japan; Petr Kopecký, University of Ostrava, Czech Republic; Serpil Oppermann, Hacettepe University, Turkey; Christian Schmitt-Kilb, University of Rostock, Germany; Heike Schwarz, University of Augsburg, Germany; Murali Sivaramakrishnan, Pondicherry University, India; Scott Slovic, University of Idaho, USA; J. Etienne Terblanche, North-West University, South Africa; Julia Tofantšuk, Tallinn University, Estonia; Cheng Xiangzhan, Shandong University, China; and Hubert Zapf, University of Augsburg, Germany.

Ecocritical Theory and Practice highlights innovative scholarship at the interface of literary/cultural studies and the environment, seeking to foster an ongoing dialogue between academics and environmental activists.

Recent Titles

Critical Ecofeminism by Greta Gaard
Writing the Earth, Darkly: Globalization, Ecocriticism, and Desire by Isabel Hoving
Ecological Entanglements in the Anthropocene edited by Nicholas Holm and Sy Taffel
Ecocriticism, Ecology, and the Cultures of Antiquity edited by Christopher Schliephake
Ecotheology and Nonhuman Ethics in Society: A Community of Compassion
    edited by Melissa Brotton
The Ethics and Rhetoric of Invasion Ecology edited by James Stanescu and
    Kevin Cummings
Dark Nature: Anti-Pastoral Essays in American Literature and Culture
    edited by Richard J. Schneider
Thinking About Animals in the Age of the Anthropocene edited by Morten Tønnessen,
    Kristin Armstrong Oma, and Silver Rattasepp
Romantic Ecocriticism: Origins and Legacies edited by Dewey W. Hall

Critical Ecofeminism

Greta Gaard

LEXINGTON BOOKS
Lanham • Boulder • New York • London

Published by Lexington Books
An imprint of The Rowman & Littlefield Publishing Group, Inc.
4501 Forbes Boulevard, Suite 200, Lanham, Maryland 20706
www.rowman.com

Unit A, Whitacre Mews, 26-34 Stannary Street, London SE11 4AB

British Library Cataloguing in Publication Information Available

Library of Congress Cataloging-in-Publication Data is Available

ISBN 978-1-4985-3358-4 (cloth: alk. paper)
ISBN 978-1-4985-3359-1 (electronic)

♾™ The paper used in this publication meets the minimum requirements of American National Standard for Information Sciences—Permanence of Paper for Printed Library Materials, ANSI/NISO Z39.48-1992.

Printed in the United States of America

Table of Contents

Acknowledgments   vii

Introduction: Critical Ecofeminism   xiii

PART I: THEORY   **1**

  1   Just Ecofeminist Sustainability   3

  2   Plants and Animals   27

PART II: ILLUMINATIONS   **47**

  3   Milk   49

  4   Fireworks   69

  5   Animals in Space   91

PART III: CLIMATES   **115**

  6   Climate Justice   117

  7   "Cli-Fi" Narratives   143

  8   Queering the Climate   161

Epilogue   181

Bibliography   189

Index   215

About the Author   223

Acknowledgments

In California where I grew up, I learned about the United Farmworkers movement when our family drove the Interstate 5 corridor through the central valley, saw the workers and the fruit stands, and I began asking questions that led me to support the "No Grapes" campaign in the 1970s as a high school student. In Washington state, I marched with PCUN (Pineros y Campesinos Unidos del Noroeste), the union whose members came up to help Washington farmworkers pressure the growers for better contracts. Each year in spring, the Skagit Valley Tulip Festival draws hundreds of thousands of visitors; the highly competitive art contest selects one artist whose work will be featured on the annual festival poster; amateur and professional photographers capture the tulips' dazzling beauty, the kids on bicycles, the Dutch windmill and the Tulip Run, but can't seem to focus on the migrant farmworkers laboring in the rain-drenched fields, often with their children beside them. In Minnesota, migrant workers labor in the sugar beet fields and in apple orchards. Whether they are indigenous Mexicans, migrant or seasonal workers, citizens or undocumented, farmworkers face many challenges: education for their children, health care for the family, substandard housing, long hours, dangerous and low-waged work—all products of institutional racism. After reading Helena Viramontes' *Under the Feet of Jesus*, one of my eco-composition students told me that in the western Minnesota town where she grew up with the sugar beets and later worked as a lifeguard, she saw many white swimmers get out of the public pool when the migrant school children came to swim. "It's no coincidence that we treat migrants like dirt," she concluded.

In Minnesota, migrant farmworkers arrive and depart seasonally, like the butterflies. But these migrants are not treated the same.

While most Minnesotans don't see the Mexican migrant farmworkers who pick up to 85% of the food eaten in the United States, the annual migration of Monarch butterflies from the Oyamel trees of Mexico's Michoacan forests to the lakes and rivers of Minnesota is eagerly anticipated, celebrated, and tracked. We can't know how many monarchs arrive and depart annually, though we do know that climate change, deforestation, industrialized agriculture, and overdevelopment have damaged monarch habitat and vitality. Labor statistics document the 20,000 to 35,000 migrant agricultural workers who are recruited annually to work in Minnesota's farm fields and food processing plants. Most are permanent legal residents of the United States from the border region of southern Texas and northern Mexico who spend April through November in Minnesota, then return home during the off-season. Both monarchs and migrants contribute to food production, monarchs through their pollination and migrants through their labor.

This book's cover image is titled "Migrant Workers," a lithograph by Minnesota's Sami-American artist Kurt Seaberg. Unlike the Tulip Festival artists, Seaberg's lithograph beautifully depicts the labor of both migrant farmworkers and monarch butterflies, framing the image with visual references to the power lines above the workers, the industrial pollutants surrounding the fields where they work, and the toxins that lurk beneath the surface, ecologically and culturally, threatening the health of humans, environments, and food systems. Above, the monarch butterfly emerges again, from chrysalis to adult, offering the possibility of health and renewal for pollinators, for laborers, and for those willing to challenge the toxins of environmental racism and environmental degradations. I am grateful to Kurt Seaberg for his ecologically and culturally attuned lithography and activism that reconnects the ethics of interspecies, social, and environmental justice. In this book, I try to do with words what Seaberg does with art: illuminate relationships among social justice, transspecies justice, and ecological justice, all rooted in whether humans conceive and perceive our self-identity as *intra-active* (Barad 2007) and *kincentric* (Salmon 2002), or whether we see our identity as separate from the rest of life, superior to earthothers, and thus free to control, remake, manipulate, burden, or destroy.

Writers flourish in the context of intellectual, activist, and collegial relationships, and I am amazed and humbled by how much of my writing was brought forward by invitation and through conversation with colleagues. Thanks are due to Annie Potts for inviting me to contribute to her volume on *Meat Culture* (2016), prompting me to explore ideas for an essay that became chapter 2. I was fortunate to be able to "test drive" that essay to attentive audiences thanks to invitations from Cynthia Belmont at Northland College in Ashland, Wisconsin, and from the students organizing a Food Justice Summit at the Claremont Colleges in California. Claire Jean Kim and Carla Freccero reached out to me with an invitation to contribute to their special issue of *American Quarterly* on "Species/Race/Sex," providing the impetus to advance my longstanding interest in the politics of breastfeeding across race and species that began in 1994, with my essay in *The Ecologist*, "Milking Mother Nature: An Ecofeminist Critique of rBGH." A colleague and friend from Tamkang University in Taiwan, Peter I-Minh Huang, extended a timely invitation that helped to separate my literary from political milk studies (the paper was becoming unwieldy at sixty pages) and carve out the essay on "Literary Breastmilk in U.S. 20th Century Fiction" for the inaugural issue of *World Literature*. These endeavors were further supported by encouragement from pattrice jones, who kept telling me, "you have a lot to say about milk," and who urged me to write out my thoughts after being deeply troubled by what I saw at the annual Fourth of July fireworks. Thanks to panel cochairs Serenella Iovino and Serpil Oppermann, "Fireworks" was compressed into a ten-minute conference presentation for a session on "Material Feminism" at the Association for the Study of Literature and Environment (ASLE) 2013 conference in Lawrence, Kansas.

Like the milk essays, the climate justice and cli-fi narratives chapters were born as joined twins, soon separated for publication. The cli-fi narratives essay came first, invited by Serpil Oppermann as a keynote for Turkey's first ecocritical conference, "The Future of Ecocriticism," held in Ankara, Turkey, in 2009. That presentation and others were first published in *The Future of Ecocriticism: New Horizons* (Cambridge Scholars, 2011). The essay was later revised and updated on the invitation of Simon Estok, who was editing a special issue of *Forum for World Literature Studies* (China) in 2014, and again for Mary Phillips and Nick Rumens, who were editing a Routledge volume, *Contemporary Perspectives in Ecofeminism* (2015). Serpil Oppermann and I cochaired a panel on "Cli-Fi Narratives" for ASLE 2015, where we felt like the grandmothers of this ecocritical genre, moderating presentations from a "next-gen" of cli-fi ecocritics like April Anson, Stephen Siperstein, and Laura Wright who were already advancing this approach. Over the course of the essay's evolution, I had the good fortune of corresponding with Dan Bloom, the blogger in Taiwan who first coined the term "cli-fi" and who has created and maintains the most comprehensive website on cli-fi literature, science, and popular culture in the world.

My work on climate justice began in 2008 with presentations for the colleges where I serve on faculty, University of Wisconsin-River Falls and Metropolitan State University in St. Paul, Minnesota. The presentation morphed from an ecofeminist perspective on the linkages among global warming, world hunger, and industrialized meat production in 2008, to an exploration of links among sexism/speciesism/climate change in 2011, and into its present form, which was continually augmented and refined thanks to audience feedback during invited presentations at SUNY-Stony Brook's Humanities Institute (thanks to Sophia Christman-Lavin for hosting and introducing the talk), the University of Kansas (thanks to Schuyler Kraus), and the University of South Dakota (thanks to Meghann Jarchow) in 2014 as well as Indiana University of Pennsylvania (thanks to Susan Comfort) and UC-Santa Barbara (thanks to Corrie Ellis and John Foran) in 2015.

Lori Gruen talked with me about the early stages of U.S. space exploration and its programs with chimps. Her longstanding advocacy for interspecies justice, her research and activism for primate freedom, and her friendship over the years have contributed inestimably to my work, particularly for material on animals in space discussed in chapter 5. Lorraine Kerslake and Terry Gifford invited me to contribute to their coedited special issue of *Feminismo/s* on "Ecofeminismo/s: mujeres y naturaleza," prompting me to explore outer space while setting a clear deadline here on earth, and Bridget Flynn kindly hosted me at Oberlin for the first "animals in space" presentation. Both Lori Gruen and Carol Adams co-organized the Wesleyan conference honoring the work of ecofeminist Marti Kheel, and that conference prompted my work on ecomasculinity and ecosexuality, writing that was later augmented when I met Beth Stephens at the Center for Twenty-first Century Studies conference at the University of Wisconsin-Milwaukee, "Anthropocene Feminism" in 2014. Beth Stephens and Annie Sprinkle's work on ecosexuality and ecoerotics as politics have energized and delighted me, particularly with their guided "Ecosexual Tour" through the Walker Art Center's exhibit, "Hippy Modernism," and I am honored that they use my ecofeminist writings as part of their theoretical foundation. Through their work, I discovered I am an ecosexual!

Along with these supportive professional, collegial, and artistic connections, it is family and community who make writing possible. I gain nourishment from the dharma talks and mindfulness practice at Common Ground Meditation Center, from the weekly sermons and racial justice activism at First Universalist Church, from participating in the annual BareBones Productions Halloween performances and the Heart of the Beast MayDay parade, as well as from the Mississippi River, and the plants and animals who share this ecoregion. At home, my writing is supported by Sasha, my feline companion, and by my daughter Flora, whose curiosity, compassion, patience and playfulness motivate me every day; she will be happy that we can now take that long-promised summer road trip, after the past three summers of writing. My mother, Beverly, has never lost faith in me over the years; from her, I learned feminism. She is the ground on which I stand.

Like most ecoactivist-scholars, I dedicate the merit of my writing to present and future generations in the hope that it will contribute toward building a just world, one more intimately attuned to life: to our interbeing with humans of diverse races, genders, nations, and with all species and elements of this one precious earth.

CREDITS

Chapter 2 was originally published as "Interrogating 'Meat,' 'Species,' and 'Plant'" by Greta Gaard in *Meat Culture*, © Brill Publishers 2017. Reprinted with Permission.

Chapter 3 was previously printed copyright © 2013, The American Studies Association. This article first appeared as "Toward a Postcolonial Feminist Milk Studies," in *American Quarterly*, 65:3 (2013), 595–618. Reprinted with Permission.

Chapter 4 was originally published as "In(ter)dependence Day: A Feminist Ecocritical Perspective on Fireworks" by Greta Gaard in *International Perspectives in Feminist Ecocriticism*, edited by Greta Gaard, Simon C. Estok, and Serpil Oppermann, © 2013, Routledge. Reprinted by permission of Taylor & Francis.

Chapter 5 was originally published as "Animals in Space: An Ecofeminist Perspective on Chimponauts, Laika, and Biosphere II" in *Feminismo/s 22: Ecofeminismo/s: mujeres y naturaleza,* (2013), 113–145. Reprinted by permission of *Feminismo/s*.

Chapter 6 is reprinted from *Women's Studies International Forum*, Vol. 49, Greta Gaard, "Ecofeminism and Climate Change," 20–33, 2015, with permission from Elsevier.

Chapter 7 is reprinted from *Contemporary Perspectives on Ecofeminism*, edited by Mary Phillips and Nick Rumens; "From 'Cli-Fi' to Critical Ecofeminism: Narratives of Climate Change and Climate Justice" by Greta Gaard, 2016, Routledge. Reprinted by permission of Taylor & Francis.

Chapter 8 was originally published as "Toward New EcoMasculinities, EcoGenders, and EcoSexualities" by Greta Gaard in *EcoFeminism: Feminist Intersections with Other Animals and the Earth*, edited by Carol J. Adams and Lori Gruen, © Bloomsbury Academic, 2014. Reprinted with permission.

Epigraph from *Environmental Culture: The ecological crisis of reason* by Val Plumwood (New York: Routledge, 200) reprinted with permission.

Epigraph from "Nature in the Active Voice" by Val Plumwood (*Australian Humanities Review*, #46. May 2009) reprinted with permission.

Figures 4.1, 4.2, and 4.3 provided by Getty Research Institute, Los Angeles (P950001).

Introduction

Critical Ecofeminism

> How can we re-present experience in ways that honour the agency and creativity of the more-than-human world?
>
> *—Val Plumwood (2007, 19)*

> Water is life. We are the people who live by the water. Pray by these waters. Travel by the waters. Eat and drink from these waters. We are related to those who live in the water. To poison the waters is to show disrespect for creation. To honor and protect the waters is our responsibility as people of the land.
>
> *—Winona LaDuke*

As a body of water, I envision the journey that culminates in this book as a series of tributaries and flows: some academic, some experiential, some pieced together by art, some dribbling off into the soil, lying latent for the future (Neimanis 2012, 2013). The activist-scholarly explorations gathered here are shaped by longstanding relationships with place and community, water and trees, earth- and animal-others, forming ethical commitments that have grown progressively deeper and more entangled. Combined with small winged seeds from the Plumwood tree, these chapters germinate a more critical ecofeminism, whose transcorporeal relations with earthothers will soon become evident.[1]

In his work articulating an ecocritical postcolonialism that places Val Plumwood's *Environmental Culture* in dialogue with postcolonial theory, Richard Watts (2008) begins with a developmental model of "three stages in the evolution of academic discourses" that can readily be applied to ecofeminism:

the first substantial sign of the constitution of a discourse is the wide recognition of a few master thinkers of its problematics and approaches; the second is the appearance of anthologies of current scholarship (typically written in an affirmative mode) and foundational texts that further circumscribe the intellectual area in question; and the third, late stage is the moment of interrogation—the simultaneously internal and external questioning of the assumptions of the field. (Watts 2008, 251)

Having authored an ecofeminist critique of the "wave" model's shortcomings for narrating histories of ecocritical and feminist theories alike (Gaard 2010a), I am well aware of the ways such models can be skewed to privilege popular standpoints, excluding minor yet important developments, perpetuating anachronisms, and overlooking synchronicity as well as confluences among diverse developments of thought. I am concerned with the ways the "wave" model implicitly focuses on Euro-Western intellectual developments, failing to describe the ways indigenous and other non-western perspectives have not only developed but also persisted to articulate deeply ecological and just perspectives. Approaches not named "ecofeminist" may still articulate views and conclusions that ecofeminists have uncovered via different intellectual lineages. Both the ecojustice conclusions and the lineages merit attention. In conjunction with these caveats, a speculative application of Watts' three-stage evolutionary model can be used to illuminate the strengths and shortcomings in ecofeminism's developments, thereby grounding my argument for advancing a more *critical ecofeminism*, a term that originates in the work of Val Plumwood.[2]

In ecofeminist discourses, first-stage foundational thinkers were largely Euro-Western, and include Susan Griffin, Karen Warren, Carol Adams, Petra Kelly, Marti Kheel, Charlene Spretnak, Elizabeth Dodson Gray, Ynestra King, Carolyn Merchant, Vandana Shiva, Maria Mies, Ariel Salleh, Val Plumwood, and Mary Mellor. These and other writers included in Leonie Caldecott and Stephanie Leland's *Reclaim the Earth: Women Speak Out For Life On Earth* (1984) named the gender/race/species/ecology/nation connections operating in hierarchical dualist thought and the logic of domination (Warren 1990), as well as the spiritual, psychological, political, philosophical, historical, economic, and activist ecofeminist standpoints from which ecofeminism developed. Anthologies and collaborations that articulated the theory's second-stage blossoming and diversity include Judith Plant's *Healing the Wounds: The Promise of Ecofeminism* (1989), Irene Diamond and Gloria Feman Orenstein's *Reweaving the World: The Emergence of Ecofeminism* (1990), Carol Adams' *Ecofeminism and the Sacred* (1993), Greta Gaard's *Ecofeminism: Women, Animals, Nature* (1993), Maria Mies and Vandana Shiva's *Ecofeminism* (1993), Karen Warren's *Ecofeminism: Women, Culture,*

Nature (1997), Carol Adams and Josephine Donovan's *Animals and Women* (1995). Amending Watt's theory, this second stage included another batch of single-authored texts and key concepts articulated by additional and foundational thinkers: Catriona Sandilands (1999) on ecofeminist democracy and citizenship, Noël Sturgeon (1997) on global feminist environmental justice, Greta Gaard (1998) on Green politics and ecofeminism, Chaia Heller (1999) on social ecofeminism and the erotic, Barbara Noske (1997) on speciesism and Marxist ecofeminism, Chris Cuomo (1998) on ecofeminist flourishing, Irene Diamond (1994) on the politics of reproduction, Susan Hawthorne (2002) on globalization and biodiversity, Sherilyn MacGregor (2007) on ecological citizenship and care politics, and Marti Kheel (2008) on vegan ecofeminism's critique of traditional euro-western environmentalisms.

Watt's third stage, the "moment of interrogation" from internal as well as external inquisitors who challenge foundational assumptions of the field, arose concurrently with the second stage via Janet Biehl's *Rethinking Eco-feminist Politics* (1991), a work that mistakenly characterized all branches of ecofeminism by one of its branches, the cultural, essentialist, and eurocentric branch. Though Biehl's hasty generalizations and straw woman arguments were exposed in multiple reviews of her work (Buege 1994, Gaard 1992, Gruen 1992, Plumwood 1992), these misrepresentations gained popularity at the same time as more reasoned critiques of essentialism were raised internally by ecofeminists such as Vicki Davion (1994) and Chris Cuomo (1992), and externally, by feminist environmentalists like Bina Agarwal (1992). Here's where the story gets muddled, as critics—from reactionary feminists such as Beth Dixon (1996), Kathryn Paxton George (1994), and Mary Zeiss Stange (1997) to an ecofeminist such as Plumwood (2000) herself—intermixed well-grounded critiques of essentialism with human-supremacist critiques of ecofeminism's posthumanist antispeciesism.[3] Certainly, Plumwood's commitment to placing a posthumanist and postcolonial interspecies ethics within her own critical ecofeminism became evident in her posthumously published (2012) articulation of "ecological animalism," but well before that, by 1996, feminist critics had already seized on animal ecofeminism as a way to disregard ecofeminism's real challenge to human centrism, and despite ecofeminists' factual and well-reasoned rebuttals, these external, humanist, and speciesist challenges carried the day—or rather, the decade (Gaard 2011).

As Watt argues, the third stage of theoretical development is critical to the vitality of the discourse: "at this point, a discourse either sustains and integrates the critiques leveled at it and evolves, or does not and ossifies, opening itself up to the charge of nostalgia" (252). Ecofeminism's continued vitality is indebted, in part, to its ability to have integrated and benefited from the critiques of essentialism and eurocentrism. Works external to ecofeminism, yet drawing explicitly on Val Plumwood's critical ecofeminist work in both *Feminism and the Mastery of Nature* (1993) and *Environmental Culture* (2002) to address intersections of race, nation, gender, and species, include Graham Huggan and Helen Tiffin's *Postcolonial Ecocriticism* (2010), and Richard Watts' own 2008 essay. Postcolonial ecofeminism has been articulated in Laura Wright's *Wilderness into Civilized Shapes: Reading the Postcolonial Environment* (2010), and diverse works exploring the intersectionalities[4] of race, gender, and species include A. Breeze Harper's *Sistah Vegan: Black Female Vegans Speak on Food, Identity, Health, and Society* (2010), Lisa Kemmerer's *Sister Species: Women, Animals, and Social Justice* (2011), Carol Adams and Lori Gruen's *Ecofeminism: Feminist Intersections with Other Animals and the Earth* (2014), and Laura Wright's *The Vegan Studies Project: Food, Animals, and Gender in the Age of Terror* (2015). More racially and culturally diverse her stories of ecofeminism have been recuperated through studies such as Niamh Moore's *The Changing Nature of Eco/Feminism: Telling Stories from Clayoquot Sound* (2015) and articles such as Susan A. Mann's research conjoining the histories of U.S. urban environmental activism with conservation and preservationist movements, demonstrating that "women such as Mary Austin, Jane Addams, Mary McDowell, Emma Goldman, Rose Schneiderman, Crystal Eastman, Ida Wells-Barnett, and Lugenia Hope were truly 'pioneers of ecofeminism and environmental justice' in that they served as advocates for both women's rights and the environmental concerns of marginalized peoples" (Mann 2011, 20). Recent international essays, special issues of journals, and full-length studies in political theory and ecocriticism confirm ecofeminism's development beyond both essentialism and eurocentrism.[5]

The work in this book advances beyond this "third-stage" ecofeminism to a fourth-stage *critical ecofeminism* through the practice of attentive listening, from the germinal and continuing trajectory of critical ecofeminism grounded in Val Plumwood's work, and from scholarly activist engagements with environmental justice, interspecies justice, queer climate justice, posthumanisms[6] (i.e., plant studies), and sustainability efforts. The Introduction and chapter 1 together develop this three-pronged fork of ideas, giving readers an implement for savoring the dishes served throughout the volume.

LISTENING

I am not talking about inventing fairies at the bottom of the garden. It's a matter of being open to experiences of nature as powerful, agentic and creative, making space in our culture for an animating sensibility and vocabulary.

—*Val Plumwood, "Nature in the Active Voice" (2009)*

Seen through the lens of foundational scholarship in feminist communication theory, human-earthother communications in Western cultures are interstructured with these cultures' patterns of gendered communication. English language communication scholars such as Robin Lakoff (1975), Dale Spender (1980), and Cheris Kramarae (1981) have documented the ways that features of women's relational talk are culturally interpreted as subordination and deference to other authorities through their use of tag questions ("it seems indigenous standpoints are being excluded, doesn't it?"), hedges ("sort of" or "kind of unscientific to omit relevant data"), apologies ("I'm sorry, but climate justice seems integral to sustainability"), and frequent interruptions. Women's gendered role in conversation requires they provide linguistic support for and stylistic accommodation to the dominant speakers, rather than leading in conversational innovation. The norm dictates that nondominant women, men, and genderqueers alike continue topics introduced by dominant speakers (usually but not always privileged men), but when women and nondominant others introduce new topics, these topics are rarely taken up.

Feminist communication scholars have looked not only at whose speech merits attention, but also at who listens. Speaking is associated with power, knowledge, and dominance, while listening is associated with subordination. Not surprisingly, feminist methodology emphasizes *listening* as a hallmark of good scholarship—listening to one's research subjects, to the oppressed, to one's activist and scholarly community—and creating structures for collaboration, whereby the research subjects can themselves set the agenda, express needs, and benefit from the scholarly endeavor (Gaard 2012).

The original focus on gender inequality in foundational feminist communication scholarship has been expanded by today's queer feminist antiracist activists, who advise white allies to listen to people of color, straight and queer allies to listen to transgender activists, and male feminist allies to listen to women. Though feminist communication scholars did not originally consider listening to more-than-humans, ecofeminists have developed this implicit component of their work. To conclude her essay on "Animal Rights and Feminist Theory," Josephine Donovan explained, "We should not kill, eat, torture, and exploit animals because they do not want to be so treated, and we know that. If we listen, we can hear them" (1990, 375). The advancement of knowledge in the sciences as well as the humanities can be measured in part by how well scholars are listening—to one another, to their communities, and to their research partners, the subjects themselves. As Deborah Bird Rose explains,

> The backgrounding of listening (receiving information) is also part of the structure of hyperseparated dualisms [described by Plumwood's Master Model]: to speak is the human prerogative (because we have language), it is the active mode of being; listening (or being spoken to) is the passive or recipient position. The

power relation is clearly hierarchical: those who speak are more powerful than those who are spoken to. I am proposing that listening, and more broadly, paying attention, should also be considered an active verb; indeed in multispecies creature communities, it must be so considered. To pay attention is to exercise intelligence, to know so as to be able to inter-act. (2013, 102)

Listening as a way of knowing and avenue for scholarship emerges from many cultural traditions. As Robin Wall Kimmerer explains in her book, *Braiding Sweetgrass: Indigenous Wisdom, Scientific Knowledge, and the Teachings of Plants* (2015), indigenous North American communities have been listening and learning the language of plants for millennia, enabling them to harvest food and to reciprocate the gifts through nourishing the land, protecting the water, and practicing gratitude. "In the old times, our elders say, the trees talked to each other," Kimmerer explains (19), and it has taken centuries for western science to discover that, in fact, "trees *are* talking to one another" (20). To hear them, "you have to concentrate; you have to give yourself to the listening" (Kimmerer 110). This rich community of indigenous, feminist, and trans-species listening theory is our birthright as earth citizens.

But most euro-western children grow up without this knowledge legacy, and must learn trans-species listening on their own. When I first started listening to trees at my grade school in Los Angeles, I frequented a grove of eucalyptus trees on the campus, nestled at the bottom of a hill dividing the upper school from the elementary school. Ivy, peeling bark, fallen leaves, and nut caps lay together, thick on the ground. Even with cars parked at the lower hillside level and parents driving the circle to pick up or drop off their children, the eucalyptus grove retained an equanimity I could sense. Many days at recess, I would enter the grove with soft footsteps to sit among the trees, breathing in their calming, minty exhalations. Sometimes my friend Mavis would join me among the trees, until our mothers ended the friendship when we asked for a sleepover: the mothers exchanged glances and refused our request, and on the drive home, I learned that I was white, and Mavis was black. The next day at school Mavis and I went to the grove and spoke quietly, examined each other's hair and hands and feet, sniffed each other's necks, and grieved the ending of our friendship. We were too young to think of challenging our mothers; that would come later. Black girl, white girl, eucalyptus trees: we all seemed out of place. Years later, I learned that eucalyptus trees were not California natives, but were colonial transplants from Australia; Mavis' and my transplantations were colonial too, and it took even longer for me to understand the entanglements among our various displacements.

Although I gave the trees my attentive presence, I was still unprepared for them to say anything. Their silence seemed linked to the invisible, pervasive silences around race.

The misconception about trees was corrected in St. Paul, Minnesota, where I took up running five miles a day to compensate for the long hours spent sitting and studying during graduate school. My route took me through a grove of bur oak trees on a grassy rise above Lake Como, down and around the lake for a couple of miles, and back up through the trees. I looked forward to this daily run for the way my body changed as I moved through the trees: it felt lighter, almost effervescent, as if the trees and my body exchanged greetings. I felt recognized, visible as an *earthother*, kin to these trees. Our brief conversation began and ended one day after a run, when I took a book to read with the trees. Their presence was so gracious that I wanted to sit and savor their company rather than just racing through. Book in hand, I entered the grove and walked rather aimlessly looking for a suitable tree trunk against which I could lean comfortably. "Over here," one tree *messaged* me with the image that "there's a curve in this trunk that will just fit your back."[7] Compliant and curious, I walked over to the tree of reference and on the other side, which I hadn't been able to see, I discovered a long vertical hollow, well formed for holding my body upright at rest. But then I ruined the moment, as my amazed thoughts shouted, "Trees don't talk!"

And in that instant, the grove of trees turned into wood.

No amount of remorse could change them back. I did not read in the "woods" that day, but began searching for studies on human-tree communications. Apart from the much-refuted work in *The Secret Life of Plants* (1973), there wasn't much scientific—or feminist—research on the topic.

Reading Deborah Bird Rose's (2013) discussion of Plumwood's philosophical animism, I learned that Plumwood had a similar experience in perceiving earthothers:

> She was able to communicate radically open ways of experiencing the world because she herself had had such experiences. She once told us a story about something she encountered when walking along the edge of a wooded area and hearing some birds singing sweetly. When she glanced up she saw that they were crows. The instant they saw her noticing them, their voices changed into the familiar raspy crow register. (107)

Clearly, Euro-Western culture is so permeated by Cartesian rationalism that children are taught at an early age not to receive—and certainly not to trust—the information being sent continuously by the animate world that surrounds us, and the diverse human communities with whom our lives are interwoven. In contrast, indigenous cultures around the world have stories of trans-species communications and even kinship, ancestry with earthothers—other species, as well as ecological entities such as plants, waterbodies, rocks and mountains, sky, and wind.[8] And while these communications cannot (yet) be wholly verified by rationalist science and technologies, our material kinship with earthothers is biological science nonetheless. These earthothers continue to send messages about their experiences through the air and through their behaviors, their flourishing or languishing, and their toxicity and demise. Who is listening?

The other part of my inherited misconceptions, race and white supremacy, was addressed in my thirties, when I began teaching at a radical environmental and transdisciplinary college, housed in a traditional but environmentally focused university in the Pacific Northwest. Immediately supported in developing an Introduction to GLBT Studies course and creating a university-wide GLBT Studies Minor, I was invited to serve on the county's Human Rights Task Force, an organization launched in response to a cross-burning near migrant farmworkers' housing. The Southern Poverty Law Center research had classified the Pacific Northwest as hosting the second most dense concentration of hate groups outside of the U.S. Southern states, a cultural climate that surely fostered the 1998 murder of Matthew Shepard, a gay youth attending college in Laramie, Wyoming.

To serve on the Task Force Board, I was invited to undergo a weekend-long unlearning racism training, where I learned the history of white colonialism and the pervasiveness of white racism in education. I learned the ways white privilege had invisibly boosted "achievements" I thought were mine alone. I learned the legacies of violence and theft, the crushing of the human spirit and the ways that white people had been trained *not to listen*, not to see, in order to preserve a system of white privilege. I learned what I had lost by serving as an unwitting participant in that system: friendships. Rich stories and histories. Knowledge beyond my white education—the indigenous agricultural practices of Central and South America, the Egyptian origins of mathematics, China's developments of waterways. I learned how colonialism and white racism had distorted all our relationships—with the land, with indigenous people, and with our ability to listen and to love.

To shatter the system of white privilege in my own life—as a foundation for working politically on the Task Force and elsewhere in the world—I would have to let go of being right, of being an authority; I would need to take risks in learning and listening, and survive getting it wrong. I would need to put myself in places where I was an outsider: attend a pow-wow, learn the land and culture of the Makah and Lummi communities where I lived, organize and march with Mexican migrant farmworkers, learn the history and diversity of Asian communities on the West coast. So I undertook this training; it is ongoing, still. Accurately listening to place meant listening to all these earthothers.

Western cultures' "not-listening" practice includes all those conceptually associated with nature, most obviously the First Nations people who are speaking their distress in the Arctic region and the Canadian province of Alberta, where oil colonialism threatens the lives of indigenous people, grizzly bears, caribou, and all life on earth. If we look at the data on environmental degradation, for example, it's hard to deny that the bees are speaking eloquently about neonicotinoid pesticides through their deaths (named "colony collapse disorder," a phrase that blames the bee colonies rather than industrial agro-chemicals), or that fish in the Athabasca River are speaking of ecosystem distress through their tumors and bodily deformations. While the messages of ecoactivists and indigenous peoples are conveyed through words, art, and activism, most trans-species communications come through images or feelings rather than words. Being open to receiving their sounds and images, I've been able to collaborate with and free a sparrow trapped in a chimney, a black-and-yellow bumblebee in a basement; I've listened to Makah women elders in Washington, and joined the activism of Anishinaabe ecojustice advocates in Minnesota. But the language of rocks took me by surprise.

In July 2014, I took a week's canoe trip in the Boundary Waters of Minnesota. My companion and I had chosen to canoe the rapids around Basswood Falls, and were taking a lunch break on a shelf of rocks that funneled the river into a rapid chute of rushing water. Not thinking anything at all, but simply observing, I was looking attentively at the river and the rocks when—the English language offers no exact words for what followed, because it doesn't recognize these phenomena—the rocks *showed themselves to me*: as I watched, the *rocks flowed*. They flowed together and apart, and I saw lava; I saw ice; I saw the cooling and settling of the rocks until their movement stilled into form, and they appeared again as rocks. Settled. What the rocks had showed me was their history, their story, their becoming. The idea of these rocks as inert and motionless was as laughable as if someone had seen me sitting there after lunch, and concluded that I would spend eternity, sitting. Clearly, Euro-Americans need to take a longer view to recognize the animacy of our earth companions. At that moment, I knew I was sitting with and upon my ancestors. My understanding had been irrevocably changed: *all rock is flow.*

This is not really an experience Euro-Americans talk about with scholars and readers whom we want to impress as rational creatures ourselves. But I did confide in a feminist colleague, who promptly sent me Jeffrey Jerome Cohen's "Stories of Stone" (2010), where he writes "Flow is the truth of stone, not its aberration. All rock is motion, all that is solid a lie" (57).[9] When another colleague sent me Plumwood's "Journey to the Heart of Stone" (2007), I knew this experience of trans-species communication was integral to critical ecofeminism.[10] There, as part of her project for re-enchanting or re-enspiriting the realm of matter—in other words, "reclaiming agency and intentionality for matter"—Plumwood argues for "the re-materialisation of spirit as speaking matter" (18). "Stones have spirit and narrative," she insists, an "energetic power and identity" (20). She describes stone as "the skeleton of our planet" (20) and describes her journey into the stone country of Arnhem Land Plateau in Australia as experiencing "the land of the Old Ones" (30).

In my ecomemoir, *The Nature of Home: Taking Root in a Place* (2007), I struggled to give voice to my mineral eroticism as a queer femme rock-climber, in passages where I searched for language to describe climbing with my buddies: "Almost all of us see vertical rock and feel desire, a movement inside our bodies toward the rock that can't fully be explained. 'It's an 'up' thing,' we say." On reaching a summit in the Wind River Mountains of Wyoming, I marveled that "the granite boulders and rocks around us reminded me of bones, as though we were sitting on the earth's own skeleton. How very alike all the elements of the earth are, and how clear it seemed there, on that summit, that we are made of the same stuff: bone, wind, sky."

In Plumwood's writing, I recognize a critical ecofeminist framework for the convergence of scholarship and embodied experiences. Here, I extend that framework in dialogue with environmental justice and sustainability studies, as well as current trends in plant studies, material feminisms, postcolonial ecocriticism and critical animal studies.

CRITICAL ECOFEMINISM

If this whole world consists of *Vibrant Matter*, as Jane Bennett (2010) argues, and has *agency* as Karen Barad (2007) and other new materialists concur, how shall we go about our lives in ways that promote the flourishing (Cuomo 1998) of all beings—at the same time as we acknowledge that all life feeds on life? This question of interbeing and ethical consumption—of our earthothers, both plants and animals—is explored in chapter 2. The volume's overarching question advances contemporary conversations about sustainability using a feminist lens to ask: How can we understand the entanglement of alienation, hierarchy and domination in terms that are simultaneously social, economic, ecological and political? And how can this understanding be used to leverage stronger and more joyful alliances for climate justice, reflecting insights, and commitments that are simultaneously feminist, queer, anticolonial, and trans-species?

To address these higher-level questions, I interrogate some very specific material throughout this book: milk (chapter 3), fireworks (chapter 4), space exploration (chapter 5), climate change realities and narratives (chapters 6 and 7), gender and sexuality (chapter 8). How has the commodification of our first human relationship, mother with infant, and its relational material, milk, affected the lives of mothers, children, and workers across differences of nation, culture, and class? Why has something like fireworks that so many cultures associate with celebration become simultaneously an instrument of material violence and affective terror? How do our early attempts at colonizing outer space build upon and reinforce hierarchies of species, race, class, and gender? Can the stories we tell about the future of climate change actually help us avert those stories of disaster coming true? Or are these fictions only preparing us to accept a future previously deemed unacceptable? And what role is there in ecopolitical movements for the gender traitors fleeing dominant heteromasculinity?

Although my past work has been conducted under various theoretical labels—from "ecofeminism" to "vegetarian/animal ecofeminism" or "feminist animal studies," and from "postcolonial feminism" to "material feminist ecocriticism"—I recuperate and continue my work here with the term *critical ecofeminism* for the ways that this ecofeminist perspective advances on the earlier findings of feminist animal activists, feminist peace and antinuclear activists, feminist environmental justice activists and queer feminist environmentalists as well as antiracist ecofeminists. Critical ecofeminism benefits from past lessons about gender and racial essentialisms, as well as from the more contemporary critical dimensions of economic, posthumanist, and postcolonial analysis. It offers helpful critiques and augmentations to ongoing conversations within environmental justice and sustainability studies discourse. It grows in dialogue with queer ecologies, delights in critical animal studies scholars' acknowledgments of their field's ecofeminist animal studies roots, and gratefully vows to inherit and extend Val Plumwood's legacy by advancing her own term, *critical ecofeminism*.

Best known for her critique of the Man of Reason and her definition of the Master Model in *Feminism and the Mastery of Nature* (1993), and the augmentation of her theory a decade later, in *Environmental Culture: The ecological crisis of reason* (2002), Val Plumwood's commitment to place influenced her decision to take the name of Plumwood mountain and its emblematic Plumwood tree, the way that some partners take their spouse's name in marriage. After her untimely death in 2008, Plumwood's Australian ecological and feminist colleagues—Freya Mathews, Kate Rigby, Deborah Bird Rose—compiled her unpublished essays into an online volume that would have been Plumwood's third monograph (2012), and wrote an introduction that stands as a lasting tribute to Plumwood's ecological feminist philosophy and actions. They also ensured that her body was given a green burial on the land she had loved for so long.

Plumwood's contributions to ecological feminist theory and environmental humanities alike are both foundational and prescient. In *Feminism and the Mastery of Nature*, Plumwood shows how the binary thinking that underlies

Western culture and pivots on the human/nature, mind/matter dualisms not only is gendered, raced, and classed, but also constructs a colonialist identity which Plumwood calls the Master Model. Drawing on developments in psychology, philosophy, economics, and political science, Plumwood describes the five linking operations that function together in creating this Master Model: backgrounding (in which the Master utilizes the services of the other and yet denies his dependency), radical exclusion (in which the Master magnifies differences between self and other, and minimizes commonalities), incorporation (in which the Master's traits are the standard against with the other is measured), instrumentalism (in which the other is constructed as having no purpose other than to serve the Master), and homogenization (in which the dominated class of others is seen as uniform and undifferentiated) (Plumwood 1993, 42–56). These value dualisms create value hierarchies that construct an artificially distinct and superior Master identity by justifying the inferiority, subordination, and colonization of indigenous people, people of color, animals, and the natural world, as well as women, emotions, and the body, as Plumwood argued; as I have argued, these value dualisms also include sexually diverse beings and behaviors, and the erotic itself.[11] Along with other ecofeminists, Plumwood emphasized that struggles for environmental health and social justice are conceptually intertwined; moreover, as long as we do not address the conceptual bases for these forms of oppression and focus only on their symptoms (as much traditional environmentalism has done), these pervasive structures of oppression will continue to operate.

But at the ecofeminist conference at the University of Montana in 1995, Plumwood parted ways with ecofeminists on the topic of species justice. Perceiving Carol Adams as articulating a universalist mandate for vegetarianism, and overlooking Deane Curtin's clear articulation of a contextual moral vegetarianism, Plumwood developed a straw woman critique of ecofeminism as "Ontological Vegetarianism," and renamed her approach as "critical feminist-socialist ecology."[12] Plumwood's later work advances "ecological animalism," an interspecies ethic that—like Curtin's contextual moral vegetarianism—challenges industrialized animal agriculture while still respecting the traditional foodways of indigenous cultures. Although Plumwood's mischaracterization of posthumanist ecofeminisms has been definitively refuted by writers such as Richard Twine (2014) and David Eaton (2002), and by postcolonial ecofeminist writings exploring food practices across cultures (Kim 2007, 2010; Gaard 2001), the allegations in Plumwood's critique perpetuated the very homogenization of ecofeminism that Plumwood deplored, damaging the theory her work had served to ground. Recuperating a critical ecofeminism in conversation with contemporary work in new materialisms, plant studies, critical animal studies, and posthumanism provides a more robust framework for articulating contextual as well as postcolonial feminist approaches to interspecies justice and culturally situated food practices.

Though Plumwood uses the term *critical ecofeminism* in both her monographs, the term is most succinctly defined in her essay on "Gender, Eco-Feminism and the Environment," where she explains "critical eco-feminism sees culture/nature (or human/nature) dualism as the key to the ecological failings of Western culture" (2004, 44). This dualism structures an "apartness emphasized in culture, religion and science" that was "shockingly challenged by Charles Darwin in his argument on the descent of species. But these insights of continuity and kinship with other life forms (the real scandal of Darwin's thought) remain only superficially absorbed in the dominant culture, even by scientists [because] the traditional scientific project of technological control is justified by continuing to think of humans as a special superior species, set apart and entitled to manipulate and commodify the earth for their own benefit" (45–46). Plumwood argues against the "dematerializing philosophers like Plato and Descartes who treat consciousness, rather than embodiment, as the basis of human identity," and advocates situating human identity "in material and ecological terms" (46). To do so, we must engage in "(re)situating humans in ecological terms and non-humans in ethical terms," for "the two tasks are interconnected, and cannot be addressed properly in isolation from each other"(57). Her debate between ecological animalism and ontological veganism concludes with the claim that "our food choices are shaped and constrained both by our social and by our ecological context" (56), cohering almost word-for-word with Deane Curtin's (1991) argument for a contextual moral vegetarianism:

> the reasons for moral vegetarianism may differ by locale, by gender, and by class. . . . As a 'contextual moral vegetarian,' I cannot refer to an absolute moral rule that prohibits meat eating under all circumstances. . . .The point of a con-textualist ethic is that one need not treat all interests equally as if one had no relationship to any of the parties . . . geographical contexts may sometimes be relevant. The Ilhamiut, whose frigid domain makes the growing of food impos-sible, do not have the option of vegetarian cuisine. The economy of their food practices, however, and their practice of 'thanking' the deer for giving its life are reflective of a serious, focused, compassionate attitude toward the 'gift' of a meal. . . . If there is any context, on the other hand, in which moral vegetarianism is completely compelling as an expression of an ecological ethic of care, it is for economically well-off persons in technologically advanced countries. (69–70)

Certainly, Plumwood's concern for earthothers is a clear part of her second monograph, *Environmental Culture*, and a major section in her posthumously published *Eye of the Crocodile*, where she challenges humans to place our-selves in the food chain as both eater and eaten. Her grief at the death of a wild wombat, her sensitive and nuanced reading of interspecies ethics in the film "Babe," and her own near-death encounter with a crocodile are all material for her critical ecofeminism. Perhaps inspired by the popularity of her crocodile narrative and its exemplary articulation of interspecies food ethics, in her later life, Plumwood turned to narrative as a suitable container for exploring ethical relations. Her construction of an "animist materialism" in "Nature in the Active Voice" (2009), her lively and listening relationships with rocks in "Journey to the Heart of Stone" (2007a), her approach to gardening as a participatory practice of food-sharing in "Decolonising Australian Gardens: Gardening and the Ethics of Place" (2005), and her ecological feminist economics which restore "Shadow Places and the Politics of Dwelling" (2008) to our awareness of consumption and production all use creative nonfiction as a genre that blends the personal and the political, a generative aim of feminist praxis. Most potent of all are her views on an ecological death and burial as a practice of returning one's body as food to the larger community of earthothers (2007b).

Plumwood's environmental justice views are implicit in her feminism, as her frequent references and reliance on indigenous standpoints affirm. Her critique of sustainability as a concept and practice are also immanent in her writings. For a critical ecofeminism to be relevant today, these threads need to be developed.

NOTES

1. The term "transcorporeal" comes from Stacy Alaimo (2010). "Earthothers" is a term from Val Plumwood (2002), who uses it to describe the animals, plants, and other life with whom we share this earth; this term includes humans as well.

2. In *Feminism and the Mastery of Nature* (1993), Plumwood uses the terms "critical ecological feminism," "critical anti-dualist ecological feminism," and "critical ecofeminism" interchangeably. At the time, "ecological feminism" was associated with the more philosophical branch, and "ecofeminism" was associated with cultural, radical, and, antispeciesist branches. Had Plumwood lived another ten years, she might have used terms such as "anthropocene ecofeminism," "critical material ecofeminism," "posthumanist and postcolonial ecofeminism"—all of these inflections are present in her work. In using *critical ecofeminism*, I honor and build upon Plumwood's intellectual, activist, and scholarly work in ways I hope she would receive as a tribute.

3. Arguments of Dixon (1996), George (1994), and Stange (1997) were refuted by Gaard (1997b), Gaard and Gruen (1995), Adams (1995), and Donovan (1995); most notably, Val Plumwood (2000) conveys both a well-grounded critique of Stange's illogic as well as a straw woman critique of antispeciesist ecofeminism, misrepresented as "ontological vegetarianism" which is "universalist," a "feminism of uncritical reversal," and "alienated" from nature.

4. This term originates with Crenshaw (1989).

5. Examples include Mary Judith Ress' *Ecofeminism in Latin America* (2006), Jyette Nhanenge's *Ecofeminism: Towards Integrating the Concerns of Women, Poor*

People, and Nature into Development (2011), Chhaya Datar's *Ecofeminism Revisited: An Introduction to the Discourse* (2011), Anne Stephens' *Ecofeminism and Systems Thinking* (2013), Greta Gaard, Simon Estok and Serpil Oppermann's *International Perspectives in Feminist Ecocriticism* (2013), Justyna Kostkowska's *Ecocriticism and Women Writers: Environmentalist Poetics of Virginia Wolf, Jeanette Winterson, and Ali Smith* (2013), Chia-ju Chang's *Global Imagination of Ecological Communities: Western and Chinese Ecocritical Praxis* (2014), Carol Adams and Lori Gruen, eds., *Ecofeminism: Feminist Intersections with Other Animals and the Earth* (2014), Mary Phillips and Nick Rumens, eds., *Contemporary Perspectives on Ecofeminism* (2015).

6. Plumwood's ecofeminism developed over a decade before the emergence of posthumanist theory, a perspective that rejects the premises of Western humanism—namely Protagoras' sense of man as 'the measure of all things,' Leonardo Da Vinci's Vitruvian man and its ideal of bodily perfection, along with the Enlightenment myth of progress and rationality. The humanist self/subject is white, male, and socioeconomically privileged; the different others, women and slaves, are excluded from the norm. Difference is at the heart of posthumanism, which places humans as but one among many, challenging both humanist and hierarchical values. Scholars identify between four (Braidotti 2013) and seven (Ferrando 2013) varieties of posthumanism. In brief, intellectual honesty requires that we recognize and acknowledge the many aspects of ecofeminism—gender, racial, queer, transspecies, and environmental justice, along with the animacy of all those seen as "nature"—that both predate and are utilized as foundational to the more formal emergence of posthumanism articulated in works such as Cary Wolfe's *What is Posthumanism?* (2010) and Rosi Braidotti's *The Posthuman* (2013).

7. I reluctantly use the term "messaged" that is now in vogue with the advent of cell phone technologies, for it does describe the brief, direct, and often unexpected transmission of messages from earthothers. There is as yet no suitable term in English for these communications among humans and earthothers because this conceptual possibility does not yet exist in this culture; only plant-plant communications are being discovered in these first decades of the twenty-first century. In my experiences of transspecies communications, messages are conveyed from earthothers through observations, images, and feelings rather than words. Such messages create an empirical challenge among humans who claim to receive competing messages, that is, "please hunt and kill me" versus "the lion wants to live unharmed." Such conflicts need to be tested using Chris Cuomo's (1998) ethic of flourishing, along with classic and contemporary advances in feminist antispeciesist ecological science studies (i.e., Harding 1993; Birke and Hubbard 1995; Neimanis 2012, 2013; Carey, et al., 2016).

8. See Robin Wall Kimmerer, *Braiding Sweetgrass: Indigenous Wisdom, Scientific Knowledge, and the Teachings of Plants* (Milkweed Editions, 2013).

9. See Cohen (2010, 2015). The colleague I first trusted with my story of rock's animacy is the Turkish material feminist ecocritic, Serpil Oppermann.

10. Thanks to British ecocritic Greg Garrard, who knew Plumwood's essay had been delivered as a keynote address for the 2002 biannual conference of Association for the Study of Literature and Environment, United Kingdom (ASLE-UK), and alerted me about this piece after hearing my own keynote address at Association for Literature, Environmen t, and Culture of Canada (ALECC) in August 2014.

11. Throughout our correspondence exchanging drafts of our current writings, Val often stuck post-it notes with additional scribbled commentary. In one three-post-it communique dated "24.10.95," Val concluded "I love the idea of 'queering ecofeminism.'" Val's support for my augmentation of her theory meant a lot to me as a young scholar, and my essay was published two years later (Gaard 1997). Catriona Mortimer-Sandilands and Bruce Erickson's *Queer Ecologies* (2010) advances this project of queering ecojustice and ecofeminisms; they define "queer ecology" as the understanding that "there is an ongoing relationship between sex and nature that exists institutionally, discursively, scientifically, spatially, politically, poetically, and ethically . . . the task of a queer ecology is to probe the intersections of sex and nature with an eye to developing a sexual politics that more clearly includes considerations of the natural world and its biosocial constitution, and an environmental politics that demonstrates an understanding of the ways in which sexual relations organize and influence both the material world of nature and our perceptions, experiences, and constitutions of that world. Queer, then, is both noun and verb . . . " (5).

12. Deane Curtin's contextual moral vegetarianism was developed in his essay, "Toward an Ecological Ethic of Care," which appeared in *Hypatia*'s special issue on "Ecological Feminism" (1991), along with Plumwood's own essay on "Nature, Self, and Gender." Overlooking this and other works providing clear evidence of diverse ecofeminist perspectives on interspecies justice, Plumwood published attacks on Adams (Plumwood 2000), renaming her own approach as "critical feminist-socialist ecology," and in her posthumously published chapter on "Animals and Ecology: Towards a Better Integration" (Plumwood 2012), where her "Ecological Animalism" emerges.

Part I

THEORY

Chapter 1

Just Ecofeminist Sustainability

Among first-world ecopolitical scholars and activists, the term *sustainability* has become a popular buzzword, albeit with ambiguous meanings. Is sustainability the same thing as "sustainable development," or the "triple bottom line (TBL)" of corporate social responsibility, as popular usage suggests? Or as environmental philosophers argue, is the meaning of "sustainability" closer to "ecological literacy," or even Aldo Leopold's land ethic?[1] *Environmental justice* has many contested meanings as well, ranging from a movement for distributive justice, a sharing of "environmental goods and bads," benefits and risks, to an argument for participatory and procedural justice. Different analyses of ecopolitical problems lead to different applied responses, from incremental reforms to cultural transformations: Should environmental justice activists work to rebalance the distribution of environmental benefits and risks across race and class, or should these risks outweigh the benefits, and prevent their production entirely? The historical branches and recent developments of ecofeminism also carry diverse meanings, evolving from earlier roots in cultural, radical, womanist, and socialist feminisms to current work in political ecology; posthumanist, postcolonial, queer, and transnational ecofeminisms now provide more inclusive and transformative analyses. Rooted in recognizing the links between human and environmental well-being, these three movements—sustainability, environmental justice, and ecofeminism— have synergistic potential for creating a broader and more inclusive movement for a just and ecofeminist sustainability.

In fact, each movement's strengths, shortcomings, and mobilized populations illuminate the others: in business, government and education, sustainability has been largely a white, male, and middle-class movement. At the level of community activism, environmental justice has been powered by people of color, focusing on race and class, with grassroots women doing much of the activism and prominent male leaders serving as spokespersons and theorists. Ecofeminism has been powered by feminists of diverse sexualities and nationalities: initially articulated primarily by Euro-western activist-scholars, ecofeminism's focus benefited from the intersectional theories of Black feminists (Crenshaw 1991; Collins 1990) and evolved to foreground intersections of gender, race, class, sexuality, species, and nation in a post-colonial, posthumanist framework. In the face of global challenges such as climate change, the movements for environmental justice, ecofeminism and sustainability may be more effective in collaboration.

INTERROGATING SUSTAINABILITY

The term "sustainability" rose to prominence through the 1987 World Commission on Environment and Development report, *Our Common Future*, wherein "sustainability" was defined as "development which meets the needs of the present without compromising the ability of future generations to meet their own needs" (Brundtland 1987). The concept of "sustainable development" was soon offered to third world countries with the promise of "catching-up development" wherein these countries would achieve techno-logical developments comparable to first world nations, all while working within the limits of their bioregion. But this illusory prospect could not be duplicated, since the affluence and technological sophistication of first world nations was created through the colonialist extraction of labor, environments, and other "natural resources" from third world countries, compensated at a fraction of their value, and further degraded by international economic institutions and trade agreements, all favoring the more developed countries. The thirty years since the promulgation of "sustainable development" have coincided with the most rapid and unsustainable consolidation of corporate control over nature, confirming critics' initial skepticism of the term "sustainable development" as an oxymoron. As one team of social scientists observes, "the pursuit of sustainable development goals has not resulted in either sustainability or effective mitigation of climate change," confirming the fact that sustainability as a "concept has failed to meaningfully change human behavior" (Benson and Craig 2014).

Some sustainability scholars are well aware of the term's shortfalls. One researcher chronicles the history of sustainability and its meanings, finding as many as seventy different definitions that range from population control to smart growth (Morris 2012). Others find that the "triple bottom line" of ecology, economy, and society is an ineffective measure of a business' sustainability because accurately defining sustainability as the progressive maintenance of the life-supporting capacities of the planet's ecosystems requires the subordination of traditional economic criteria to criteria based on social and ecological values, and this raises the question of whether business decision makers operating within the constraints of a capitalist system are capable of making sacrifices of profit to protect resources and ecosystems for future generations and other species. (Milne and Gray 2013, 16)

Although major texts such as *Sustainable Capitalism: A Matter of Common Sense* (Ikerd 2005) and *Capitalism As If the World Matters* (Porritt 2007) argue that public policies can be used to impose social equity and ecological integrity upon capitalist economies (Ikerd), or that Green politics can transform capitalism and make it more sustainable (Porritt), other scholars exploring sustainable development and entrepreneurship insist that "sustainability is fundamentally at odds with the prevailing model of capitalism and its emphasis on unbridled growth" (Hall, Daneke, and Lenox 2010). As Lynn Chester incisively concludes, while some find attractive the notion that capitalism may deliver sustainability, this notion is also

at considerable variance with history and reality. Air and water pollution, deforestation, desertification, soil erosion, biodiversity loss, and global warming dominate the ecological legacy from the conjunction of twentieth century capitalism's widespread use of fossil fuels, technological change, industrialization, mass production and mass consumption, and globalization. This is capitalism's relation to nature. (Chester 2010, 5)

Instead of providing accuracy, sustainability measures such as the Global Reporting Initiative, benchmarking, and the TBL of "people, planet and profit" are now "achievable in a manner which offers little or no challenge to business-as-usual," but rather function as empty signifiers in which the true meaning of sustainability gets lost (Milne & Gray 18). In effect, "the TBL may be better understood as an organizational and institutional barrier to develop ecological literacy and a fuller take-up of sustainability" (Milne & Gray 24).

Companioning the reluctance to investigate a postcapitalist ecological economics with the rise of "sustainable development" for third world countries and global corporations alike, the academy has also witnessed the rise of "sustainability studies" and the emergence of academe's flagship organization of sustainability, the Association for the Advancement of Sustainability in Higher Education (AASHE). Shortly after its formation in 2005, AASHE began offering faculty workshops in Sustainability, training higher education faculty to "uncover" the sustainability issues inherent in every discipline across the curriculum. On its website, AASHE "defines sustainability in an inclusive way, encompassing human and ecological health, social justice, secure livelihoods, and a better world for all generations." The model for sustainability in these faculty training workshops advances the inseparability of economy, ecology, and society—variously imaged as the "three-legged stool," "triple-bottom line," or a Venn diagram of three intersecting circles. One especially detailed model illustrates the interlocking circles, clarifying the intersections of any two of these elements is necessary but not sufficient to be termed "sustainability."[2] Nonetheless, workshop leaders for AASHE's "Sustainability Across the Curriculum" faculty trainings are cautioned to take a "hands-off" approach to these trainings, to acknowledge that faculty themselves are "the experts" and to "hold the term lightly" as "there will be a diversity of definitions around the term 'sustainability.'" Indeed, at the 2015 AASHE Sustainability Leaders Workshop held at Emory University, participants were given testimony from past workshop participants, one of whom asserted that "as soon as I heard there was no one right way to define 'sustainability' I was able to relax and participate." Far from the feared fundamentalism of the "one right way," clearly defining sustainability as a specific response to the ecopolitical crises of today would guide AASHE and other sustainability-focused organizations toward enacting more targeted, accountable, and effective programs and policies. So, what's preventing AASHE—the leading sustainability organization in U.S. higher education—from developing clearer definitions of sustainability for widespread usage?

On the surface, it would appear AASHE avoids mandating a single, specific sustainability definition in favor of a more democratic method, supporting its members and member institutions in defining and assessing their own sustainability goals. To assist in that assessment, AASHE has developed a Sustainability Tracking, Assessment and Rating System (STARS) for self-reporting, whereby campus representatives can evaluate their own campus' progress toward sustainability on the basis of academics, engagement, operations, and planning and administration. The idea is that universities can lead the way to sustainable practices by demonstrating the widespread appeal, utility, economic efficiency, and intellectual benefits of such practices; after experiencing these practices on campus, students at these colleges may be inspired to continue sustainability practices in their personal lives, and to disseminate these ideas at their future workplaces. But the campus sustainability data provided annually through AASHE's popular STARS relies entirely on members' self-reporting; AASHE publicizes its members' reports, but does not assess them. As a consequence, multiple and conflicting articulations of sustainability are advanced, and the overarching definition of sustainability shows little progress in advancing an intersectional approach to sustainability over "old-school" definitions of sustainability as simply another word for "environmental sciences."

In a 2013 self-study of topics addressed in the AASHE Bulletin—a publication described as representing "a sample of what is happening in the higher education sustainability community"—the data collected shows that in 279 articles, the leading topic was energy in campus facilities for AASHE member institutions, emphasizing an environmental science and technology approach to sustainability; cultural diversity and inclusion was addressed in only 55 articles, roughly a 5:1 ratio.[3] The omissions of social justice from the envisioned balance of ecology/economy/society are not restricted to AASHE's sustainability-across-the-curriculum trainings but are evident in the assemblage of Sustainability Studies programs at large. In 2015, AASHE's Academic Program database contained 1447 sustainability-focused academic programs at 476 campuses in 66 states and provinces.[4] Of these sustainability-focused programs, there are

- Associate Degree Programs (33) in Agriculture, Architecture, Building Facilities, Development, Energy, Landscaping, Technology, Watersheds, Wind Turbines;
- Baccalaureate Degree Programs (428) with Science and Technology emphases, with a few combining "Environmental and Sustainability Studies";
- (1) program in "Environmental Sustainability and Social Justice," offered at San Francisco State University;
- Master's Programs (469) emphasize sustainable development, resources, management, and technologies, with fewer programs in environmental studies, education, communication, and public administration;
- Ph.D. programs (104) in environmental sciences, law, natural resources, agriculture, development, engineering, and policy; and
- Joint Degree Programs (34) either JD/PhD or JD/MS.

While most sustainability studies program statements express a verbal commitment to implementing interdisciplinary (some even say *transdisciplinary*) approaches to sustainability, these sustainability statements all stem from disciplines in the environmental and social sciences—geography, environmental studies, sustainability studies, business, economics—and overlook the tools and contributions of the environmental humanities, the interdisciplinary nexus producing more transformative approaches to ecojustice (LeMenager and Foote 2012). In practice, at AASHE member institutions, "sustainability" often means greening technologies of development; but on a finite planet, such "sustainability" is not sustainable.

Sustainability scholars themselves have noticed that most commonly cited definitions of sustainability fail to mention economic justice and racial equity, thus distancing sustainability from the environmental justice movement, despite the fact that the two movements emerged around the same time: only four years separate the Brundtland Commission's 1987 definition of "sustainable development" from the 1991 Principles of Environmental Justice. As a "top-down" phenomenon emerging from "international processes and committees, governmental structures, think-tanks and international NGO networks," sustainability is more future oriented, while the environmental justice movement is a "bottom-up" response to local and immediate struggles for just transport, community food security, and sustainable cities (Agyeman, Bullard, and Evans 2002). But the term "environmental justice" is not often used among sustainability professionals in academe; why not? Is it because the relationship between environmental justice and environmental sustainability is not one of interchangeability, and may even be one of competing emphases? As some have argued, environmental justice can favor social justice over environment, while environmental sustainability tends to take a resource management approach (Margolis 2005). Or is environmental justice backgrounded in sustainability workshops and practices because environmental justice is so clearly political in its challenge to white privilege itself as antienvironmental—and because the majority of sustainability professionals are white? (Taylor 2014).

Environmental justice scholar Julian Agyeman has led the way toward efforts to bridge the two movements through the concept of a "just sustainability"—as opposed to a purely *environmental* sustainability—and has even keynoted an AASHE conference in 2010, introducing this intersection to an audience of sustainability educators and businesspeople, though this intersection has yet to catch on. Yet in public life, sustainability scholars researching sustainable development have uncovered "social injustices associated with some sustainability initiatives [that] are in many cases intentional" (Pearsall, Pierce & Kruger 2012, 936).[5] According to geographers reflecting on research presented at the 2010 Association of American Geographers annual meeting, allegedly "sustainable developments" ranging from New York to South Carolina and the Galapagos Islands have tended to privilege green consumerism while "compromising (rather than just overlooking) social justice concerns" (Pearsall, Pierce & Kruger 937). Most important, these scholars challenge the very definition of sustainability, asking "what is to be sustained, by whom, for whom?" and concluding that

> For some in the US environmental justice movement, the 'sustainability movement' is merely a renaming of the 'old' environmental movement which did not hire minority staff, nor take up 'door-step' or environmental justice issues, preferring instead, wilderness, resource and other 'green' issues. And in many respects it is. (Agyeman & Evans 2004, 156, 162)

This concern is shared by sustainability studies scholars in higher education, who note the proliferation of sustainability programs and the absence of comprehensive evaluation frameworks that would ensure the promised integration of ecology, economy, and society—thus, a *just* sustainability—is fully enacted across the curriculum and weighted equally in importance with decisions affecting these institutions' facilities and community relations (McFarlane and Ogazon 2011; Mader, Scott, and Razak 2013; Koehn and Uitto 2014).

Silences about race are not uncommon to sustainability conversations in academe, and are companioned by silences around class, gender, sexuality, and species. In the AASHE literature on sustainability, there is little attention to the *intersectionality* of gender, race, class, sexuality, species, and climate justice—this despite the affirmation from AASHE's 2013 Annual Report that "AASHE defines sustainability in an inclusive way, encompassing human and ecological health, social justice, secure livelihoods, and a better world for all generations."[6] As feminists and ecofeminists have repeatedly demonstrated, paradigms omitting or backgrounding discussions of race, class, gender, sexuality, and species effectively ensure that these paradigms will be marked by the "unmarked" dominant group—white, male, middle class, heterosexual, and human animals. For "sustainability" to reach its full potential, its advocates will need to recast sustainability in dialogue with an ecofeminist, environmental justice framework.

ENVIRONMENTAL JUSTICE

Today the stories of Alberta's tar sands excavations, and the devastating impacts on forests, water, wildlife and native people dominate the environmental news. We are well past peak oil, as the effort to extract, transport, and refine tar sands crude produces less oil at greater costs than ever before: 4 tons of sand and soil are used to produce 1 barrel of tar sands oil, at a rate of 400 million gallons of water per day dumped as toxic waste.[7] Increases of 30% and more in rates of rare cancers among Fort Chipewyan residents, decreases of more than 70% among caribou herds, and dangerously high arsenic levels in muskrats, ducks, and moose are diminishing the lives of indigenous people, animals, and ecosystems around Alberta's tar sands (Nikiforuk 2010).[8] Hazardous air and water pollution from tar sands operations concentrates at Fort McMurray, another home of First Nations people, and extends to nearby Edmonton, bringing along polycyclic aromatic hydrocarbons (PAHs) and other carcinogens such as benzene and styrene, all causing elevated leukemia rates for people living near tar sands oil-processing facilities. The climate change effects of tar sands have increased 21% since 2010, already warming the planet well beyond the two-degree Celsius limit scientists have set to maintain the world we have known (McKibben 2010).[9] Yet Canada's oil companies continue to seek ways to transport tar sands crude to refineries. First, it was the Keystone XL pipeline through the central United States, but numerous protests succeeded in stalling presidential approval of that line. Activists in Nebraska—ground zero for the Ogallala aquifer—built a "Cowboy and Indian Alliance" of indigenous and Euro-American environmental justice activists to lead the national resistance movement. Meanwhile, the Canadian corporation Enbridge is expanding its Alberta Clipper and Sandpiper lines across the upper Midwest, respectively promising to ship 800,000 and 600,000 barrels of crude oil per day (Fesher 2015).[10] With a history of "spills" yet to be "cleaned up"—most notably Enbridge's 2010 pipeline spill of 800,000 gallons in Kalamazoo, Michigan—and with massive local opposition in Canada and the United States, these pipelines nonetheless receive approval from U.S. state environmental agencies to cross sovereign native lands, wild ricing areas, fragile forests and wetlands. From Canada's tar sands or North Dakota's Bakken oil fields, pipelines transport crude oil across the Upper Midwest to refineries in Minnesota and in Superior, Wisconsin—where some claim there is a movement to ship this crude oil by tankers across Lake Superior, the world's largest freshwater lake. Additional pipelines run the length of North America's west coast, and from the Pacific Northwest to the Gulf of Mexico.

From an environmental justice perspective, the colocation of oil production on indigenous lands or within communities of color is no coincidence. Along with Alberta's tar sands and their impact on the First Nations communities nearby, oil refineries in Michigan disproportionately affect African-American communities. At the June 6, 2015, Tar Sands Resistance March in St. Paul, Minnesota, activists such as Emma Lockridge from Detroit, Michigan, arrived to tell stories of oil and environmental injustice, this time in the African-American community of Boynton, a Detroit suburb. There, residents are experiencing alarming rates of kidney cancers and cancer mortalities linked with prolonged benzene exposures emanating from Marathon Petroleum Corporation's tar sands oil refinery. When Marathon planned its refinery expansion, it offered homeowners in the Oakwood Heights neighborhood of zip code 48217–also shared with Boynton—a buyout that included both a base price plus 50% of the appraised value, or a minimum of $50,000 for owner-occupied homes (Halcom 2012). Not surprisingly, nine out of ten property owners were interested in the buyout, while across the way, other Boynton residents received no such offers, and saw their property values plummet to $16,000 or less by 2014, as a result of the refinery expansion (Lewis 2014). The complicity of industry and government became even more suspect in May 2013, when the Michigan Occupational Safety and Health Administration announced it was giving Marathon a safety award one week after the refinery had an explosion. In 2014, in an effort to compel the

Environmental Protection Agency to fulfill the promises of the Clean Air Act, Earth Justice filed a Community Impact Report Addendum on "The Toll of Refineries on Fenceline Communities," profiling ten environmental justice communities who share the commonality of elevated cancer and asthma rates associated with their proximity to oil refineries.[11] Citizens in many of these communities have appealed to courts and federal and state regulatory authorities for justice, but their claims are often disregarded.

Detroit, Michigan, is one of those ten environmental justice communities, with an 82.7% African-American population, and 38.1% of residents living below the poverty line. Though Marathon's oil refinery has persistently violated the Clean Air Act, faced numerous penalties, and paid more than $4.5 million in formal enforcement actions, it was still allowed to expand its capacity and take on tar sands oil processing. Seventy miles north of Detroit, Flint has become well known for its drinking water crisis which surfaced in 2014, as the nation learned Flint officials had changed the source of Flint's water from Lake Huron to the polluted Flint River, in an effort to save the city $5 million over two years: in practice, the city's economic savings was achieved at the expense of health for between 6,000 and 12,000 children now exposed to lead through their drinking water.[12] Cartoonist Matt Wuerker's (2016) depiction of the water crisis as two separate drinking fountains—one providing clear water for whites and the other, brown water for people of color—illustrates the Flint water crisis' longstanding roots in U.S. racial segregation.

Across the nation from Detroit, Hispanic and Latino residents on the north side of Corpus Christi, Texas, are exposed to the toxic emissions of six large oil refineries, which together in 2012 released over 1.5 million pounds of hazardous air pollutants including benzene, diethanolamine, and xylene. These emissions are public knowledge, yet the courts do little to protect the residents or to enforce environmental regulations.[13] And while the U.S. Environmental Protection Agency's environmental justice website claims the Agency ensures "the same degree of protection from environmental and health hazards and equal access to the decision-making process to have a healthy environment" for all communities, its record of enforcing punitive actions and providing reparations for environmental injustices falls short of these stated goals, to say the least.[14]

The environmental justice framework does well to explain these practices as environmental racism and classism. Traced back to initial resistance in Warren County in 1982, the U.S. environmental justice movement took shape in 1991 at the First National People of Color Environmental Leadership conference, where the 17 Principles of Environmental Justice were formulated. Subsequent conferences in 2002 and beyond expanded the definition to address climate justice as a manifestation of environmental justice, and to address economic injustices as integral to environmental justice: as Robert Bullard states, "We are just as much concerned with inequities in Appalachia, for example, where the whites are basically dumped on because of lack of economic and political clout and lack of having a voice to say 'no'."[15] But how does the environmental justice framework address such incidents when they occur outside the framework of an oppressed race and class?

In 2005, for example, Minnesotans were alarmed to learn that the state's flagship corporation, 3M (Minnesota Mining and Manufacturing), had for five decades discharged an annual and unpermitted 50,000 pounds of PFCs (perfluorinated chemicals) into the Mississippi River and contaminated the groundwater in Cottage Grove, a community that is 86% white, with a median household income of $81,622. Instead of listening to lead scientist Fardin Oliei, a Minnesota Pollution Control Agency's (MPCA) employee of 16 years, the Agency silenced her, prompting Oliei to file a complaint with Public Employees for Environmental Responsibility, a Washington, D.C.-based nonprofit that specializes in providing legal assistance to government whistleblowers. Two years later, in an out-of-court settlement, the MPCA gave Oliei $325,000 to drop her whistleblower lawsuit, even while the health effects on the surrounding communities—the people, water, and wildlife—remained uncertain.[16] Existing research showed that persistent exposure to PFCs bioaccumulates in human and animal tissues, increasing the risk of tumors in the liver, pancreas, and testes; adversely affecting animal immune systems; increasing the risk of prostate cancer, cerebrovascular disease, and diabetes; and producing reproductive abnormalities in offspring.[17] These and other data formed the foundation for Minnesota's Attorney General to file a lawsuit against the 3M Corporation on behalf of the people of Minnesota, 3M's remediation efforts of over $100 million notwithstanding (Mosedale 2006; Edgerly 2005).[18]

From an environmental justice standpoint, it's clear that while dumping on inner-city communities of color, indigenous lands, or rural poor communities has become almost commonplace through terms such as "sacrifice zones," illustrating environmental racism and classism, dumping on white middle-class communities gets the attention of the media, the judicial system, and the state. It prompts extensive cleanup efforts for the middle-class communities, but still does not deter corporations' illegal, antiecological and unjust practices. Clearly, this dumping isn't environmental racism or classism, so, what is it?

Forty-six years apart, two separate incidents in a wealthy community help to name the phenomenon. The first incident occurred in Santa Barbara, California, on January 28, 1969, when a massive eruption of crude oil leaked from Union Oil's Platform A into the Santa Barbara Channel, covering the entire city coastline (along with Ventura and Santa Barbara county coastlines) with a thick layer of crude. Santa Barbara's largely upper-class residents immediately deployed their resources, time and contacts with national and international elites to bring attention to the situation. Supported by widespread media coverage, Santa Barbarans held rallies, wrote letters to key Congressional officials, introduced legislation to ban offshore oil drilling, and filed lawsuits against the oil companies and the federal government. Two bird-cleaning centers were established to cleanse the oil from damaged wildlife, as seabirds either suffocate from the oil, become unable to fly, or ingest the oil on their feathers through continuous preening. Dead porpoises and whales washed up on the beaches; unusually large numbers of dead sea-lion pups were sighted on the Channel Islands, beyond the Santa Barbara coast. Equally poignant were the frantic and largely futile efforts of local citizens to speed the cleanup of beaches by dumping straw over the oil-soaked sands and then raking it up, or pouring cat litter on the sands in the hopes that it would "clump" the oil (Molotch 1970; LeMenager 2014).[19] But these new citizen activists discovered the alliance between government, corporations, and science was stronger than their local, albeit wealthy community. As Santa Barbarans experienced the inaction of federal regulatory agencies—from the Department of the Interior to the U.S. Army Corps of Engineers, the U.S. Geological Survey, and the Federal Water Pollution Control Administration—their responses moved from indignation to disillusionment and even "radicalization," according to Harvey Molotch, then a sociology professor at UC-Santa Barbara. A testament to the visibility and impact of environmental hazards when they take place in economically well-off white communities, the Santa Barbara oil spill of 1969 is credited with launching the first Earth Day in 1970, along with a raft of environmental legislation enacted throughout the 1970s by a growing environmental movement.

Why, then, almost fifty years later, on May 19, 2015, did another crude oil spill occur in Santa Barbara County? This time the rupture occurred onshore from a corroded Plains All American pipeline (with no automatic shut-off valve) bringing oil along the coast to a pumping station, where the crude is moved inland. Once again, effects on wildlife were recorded in agonizing pictorial detail: more than 100,000 gallons of oil along the coastline produced oil-soaked brown pelicans, sea lions, elephant seals, dolphins, and smaller creatures such as crabs, snails and fish, and will have lasting effects on the reproductive health of these animals. Even so, the initial flood of national news coverage—always anecdotal, rarely analytical—subsided after a few weeks into what Molotch calls "the routinization of evil" that habituates the news-watching public to further inaction and acceptance of the corporate status quo: pollution becomes routine, it is cliché that politicians are corrupt, compromises are required, and if "about half" of the oil gets recovered this is dubbed an "improvement" (Molotch 1970, 140–41).

Environmental justice analyses don't explain the impact on diverse animal species, or the repeated occurrences in areas of wealth for these Santa Barbara oil spills as coherently as an Interior Department memo from the 1969 event. Justifying the policy of refusing public hearings prefatory to oil drilling, an engineer wrote to assistant secretary of Interior J. Cordell Moore, "we preferred not to *stir up the natives* any more than possible" (Molotch 1970, 139, italics mine). The reference to colonialist relations suggests that oil extraction and transport, along with its hazardous effects on indigenous people, wealthy white people, more-than-human-animals and ecologies is an updated expansion of colonialist practice, manifesting through the corporate takeover of local, state, and national governance.

Consider Alberta's tar sands as a case in point. Before the recent tar sands boom, Alberta's primary economic industry was cattle, and before that, fur. As Andrew Nikiforuk (2010) explains, seventeenth- and eighteenth-century European colonists began by exploiting the Athabascan forests, wetlands, and wildlife in the fur trade. Whereas indigenous people traditionally lived by a communal sharing of food, the fur trade introduced money and incentivized accumulation of surplus furs for trading, eventually encouraging the rapid decline of the beaver population. Previously communal hunting grounds became divided, introducing the concepts of territorial ownership and Catholicism—thus replacing the spiritual immanence and animism of indigenous cultural views with a mechanistic, transcendent, hierarchical, and patriarchal worldview. Together with alcohol and European diseases, these forces decimated the native communities.[20] From an anticolonial ecofeminist perspective, the domination of "nature" others—indigenous people, animals, land and ecosystems—is intertwined with the construction of the Master Self (Plumwood 1993, 1997).[21]

Colonialism of earthothers continues today through tar sands operations, fracking, and industrial animal agriculture. Research such as the report by the Food and Agricultural Organization of United Nations, *Livestock's Long Shadow* (Steinfeld et al. 2006), T. Colin Campbell and Thomas M. Campbell's *The China Study* (2006), the National Academy of Sciences research on industrial animal agricultural and climate change (Springmann et al. 2016), along with documentaries such as *Forks Over Knives* (2011) and *Cowspiracy* (2014) all confirm the ways that the western industrialized animal-based diet harms human health and livelihoods, produces immense animal suffering, and accelerates climate change through methane emissions, deforestation and waste. Using vast tracts of land to feed animals for human consumption creates real material hunger for humans and wild animals, and suffering for the billions of animals instrumentalized in these animal food industries. It also creates ill health for the workers who dismember, process, and/or consume these tortured animal bodies. Neither sustainability studies nor environmental justice offer a framework for recognizing the linked environmental injustices variously harming nonhuman animals, indigenous people, and wealthy communities, a silence that suggests intersecting sustainability's middle-class appeal and environmental justice's analyses of race and class with a critical, anticolonial ecofeminist perspective may be both more descriptive and more effective in mobilizing wider cross-sections of activists.

ANTICOLONIAL ECOFEMINISM

Taking a short view of history (rather than the long view advised by our rock ancestors), western culture "forgets" its conquest of indigenous communities and nations, its occupation of Mestizo lands, and its enslavement of African-Americans. This amnesia prevents white middle-class and wealthy citizens of first world nations from recognizing oil colonialism, and from recognizing themselves as categorized on the other side of the nature/culture dualism, along with the colonized "natives." Building on centuries of colonialist explorations and conquest that began with the fifteenth-century "voyages of discovery," global corporations have accumulated the power of colonial empires, aided by post-World War II international trade agreements such as the General Agreement on Tariffs and Trade (GATT), the North American Free Trade Agreement (NAFTA), the Trans-Pacific Partnership (TPP), and international organizations such as the World Bank, the International Monetary Fund and the World Trade Organization (Klein 2014). When a Canadian tar sands oil pipeline company uses another nation's "eminent domain" laws—originally crafted to declare, claim, and compensate landowners for sites deemed as necessary for their own community's "public good"—and seizes that nation's lands for multinational corporate profits, oil colonialism is at work, advancing the theft of land and life from indigenous communities that began with the European invasions of North and South America, Africa, Asia, and the Pacific and Caribbean Islands.

As Marti Kheel has argued extensively in illuminating the interstructuring of sexism, speciesism, racism, and classism through terms like "sacrifice"—historically used to legitimate ritualized killing of nonhuman animals, young girls, and slaves[22] in order to propitiate an allegedly angry god and save the larger community—elite citizens rationalize environmental injustices enacted against animals and indigenous, third world, and/or impoverished communities of color in the first world as "sacrifice zones" (Kheel 2008). "Behind the sacrifice of animals at the altar of science" and environmental injustice, writes Kheel, "lies the ancient and tragic belief that somehow, if animals [and subordinate earthothers] are killed, [elite] human beings will be allowed to live" (Kheel 1989, 104). Sacrifice is effectively a nonreciprocal, imbalanced and instrumentalized relationship between privileged and subordinated groups, a relation that many elites have accepted as "unavoidable" in extracting resources and labor from other species, environments, and people of third world countries (as well as from the third world within the first world). But "sacrifice" reinforces illusions of safety even in the presence of material danger. To keep these illusions in place, distance is created between the sacrificed and the saved: Val Plumwood calls this operation "remoteness."

According to Plumwood, "remoteness disturbs feedback and disrupts connections and balances between decisions and their consequences" and thus an important corrective in ecological decision-making would be to "minimize the remoteness of agents from the ecological consequences of their decisions" (2002, 72). As in Plumwood's Master Model, remoteness is described along five linked operations: *Spatial remoteness* involves living somewhere remote from the places and people affected by the ecological consequences of decisions; *consequential remoteness* means the consequences of decisions fall systematically on some other person or group, leaving the decision-maker unaffected; *communicative* and *epistemic remoteness* refers to the poor or blocked communication with those affected, thus weakening knowledge and motivation for repairing ecological relationships; *temporal remoteness* involves being remote from the effect of decisions on the future; and *technological remoteness* produces well-being in places of prominence and privilege, while disregarding "waste" places conceived as "externalities" (2002, 72–73). Specifically, communicative and epistemic remoteness appears in "a society's incapacity to heed speech—warning or distress signals—from below in human society and ecological warning signals from non-human nature" (73). Plumwood argues for "minimizing remoteness" in decision-making, so that "those who bear consequences . . . have a proportionate share in the decision-making . . . sharing consequences and risks" (73). These transcorporeal and interspecies inflections for concepts of *sacrifice* and *remoteness* helpfully augment environmental justice analyses, bringing forward views that are compatible with the environmental justice framework (Pellow 2014). For the environmental justice vision to become more descriptive of current environmental injustices, its focus on race and class needs to be augmented to address the already-present elements of gender, sexuality, and species (Pellow 2016); its focus on economics can more explicitly address the ways that economic structures are gendered, and the ways that corporations enact colonialism.

Efforts toward this reframing have been underway for two decades. In 2009, Noël Sturgeon shifted her analytical framework from ecofeminism to "global feminist environmental justice," developing a theory that brings forward the feminism implicit in environmental justice, as environmental justice movements are largely originated and powered by women; however, Sturgeon's conceptual framework does not address species justice

(Sturgeon 2009). Graham Huggan and Helen Tiffin's *Postcolonial Ecocriticism: Literature, Animals, Environment* (2010) advances a strong analysis of colonialism as it has affected animals, environments, and third world communities, but lacks the feminist perspective on colonialism offered by works such as Andrea Smith's *Conquest: Sexual Violence and American Indian Genocide* (2005). Notably, David Pellow's *Total Liberation: The Power and Promise of Animal Rights and the Radical Earth Movement* (2014) brings the dimension of species justice into dialogue with environmental justice, interstructured along with the antiracism of many white radical environmental activists. But the fact that animal agriculture is one of the top three emitters of greenhouse gases (GHGs), and thus must be brought forward into definitions of sustainability and divestment, has not been addressed or linked with ecofeminism—until recently.

In 2016, critical animal studies scholar-activists met in Australia to compose "The Sydney Declaration on Interspecies Sustainability," arguing that "a key source for an enriched understanding of sustainability is ecofeminism and its suggestion that sustainability should not be discretely boxed as only a concern for the 'environment.' A sustainable relationship to the environment is linked to care and justice for other animals, women, people of colour, queers, and other 'others'" (Probyn-Rapsey et al. 2016, 113–114). As evidence of human justice concerns, the Declaration cites the "millions of Indigenous peoples and peasant farmers whose land has been stolen in processes of 'agricultural dispossession' through 'land-grabbing,'" along with "the appalling psychological and physical costs to workers in CAFOS (Concentrated Agricultural Feeding Operations) and slaughterhouses" and "the associated increases in sexual and domestic violence inflicted on their families and neighbors" (121). The Declaration authors enumerate the health benefits of plant-based diets, including "increased longevity and a reduced risk of obesity and chronic diseases including cardiovascular disease, type 2 diabetes and some types of cancer" (123). In sum, they argue that interspecies ethics are not a "private matter" of "choice" but rather a public matter of social justice (125), and propose a definition of interspecies sustainability that updates the Talloires Declaration (1990) signed by over 350 university presidents and chancellors in over 40 countries, thus providing an enhanced definition of campus sustainability: "When we focus on animal agriculture, not only in terms of GHG emissions, but comprehensively in relation to failures of social justice, including interspecies ethics, it becomes clear that socially responsible sustainability begins where animal exploitation ends" (137).

To achieve their implicit and professed goals, the environmental justice and sustainability movements must be brought into dialogue with the insights of critical ecofeminism, by interrogating the terms on which a *just sustainability* rests: ecology, economy, social justice.

IT'S CRITICAL: A JUST, ECOFEMINIST SUSTAINABILITY

In the northern hemisphere's summer of 2015, Shell's oil rigs were moving to the Arctic Sea in preparation for expanded oil drilling. Despite overwhelming evidence of climate change accelerations brought about by the burning of fossil fuels; despite calls from international panels of scientists (including Nobel Laureates) for a moratorium on oil drilling and developments; despite massive protests in the ports of Seattle and Portland, where activists in kayaks or dangling from ropes on bridges have attempted to block Shell's oil rigs from passage; despite the fragility of the Arctic ecosystem, and the documented history of "spills" which have cost millions in still-unfinished cleanup efforts, and continue to have death-dealing impacts on wildlife, water, and native communities; despite the 75% likelihood of another accident from this new drilling operation; despite widespread citizen opposition, voiced through the communicative channels of letter-writing, phone calls, and visits to elected representatives—Shell had already invested millions to expand its Arctic drilling operations, and needed only the approval from U.S. President Obama to proceed. Ignoring the rationality of all the preceding arguments, Shell's rationale was simply market-driven: the Arctic contains oil "resources" that Shell can extract and sell for billions in economic profits. The costs to the Arctic ecosystem, the local multispecies community, and the global planet are *backgrounded* in these profits.

Both Arctic drilling and industrial animal agriculture exemplify Plumwood's *remoteness* concept, and her explanations of its related failures of reason and ecological self-awareness, as well as the participatory and economic democracy that Schlosberg (2007) argues is foundational to enact environmental justice. The corporate executives and elected decision-makers who approve this drilling or factory farming, like the consumers for whom the oil and meats are marketed, are spatially and temporally remote from the violent consequences of their actions, now and in the future. Their homelands and livelihoods are not destroyed; their health is viewed as separable from the production of commodities they consume. They are also communicatively remote from the citizens and stakeholders they allegedly represent, and their failure of *listening* produces a failure of knowledge, an *epistemic remoteness* in Plumwood's terms: "The close connection between remoteness and bad decision-making . . . [illuminates] the political patterns that make some places better at the price of making other more distant places ecologically worse" (Plumwood 2002, 73). Exemplifying a national deficit in sustainability, the United States (like other western industrial-capitalist nations) is "a civilization which lacks or underdevelops ecological rationality, which sets in motion massive processes of biospheric and ecological degradations which it cannot respond to or correct, [and which] *does not match its actions to the survival aims it may be assumed to have*" (Plumwood 2002, 68, italics mine). These actions are also a failure of self-awareness, forgetting our ecological interidentity and transcorporeality. Under Plumwood's "rubric of rationality," to achieve sustainability, political agents should demonstrate a "match between means and ends" (69).

Despite their professed aims of interdisciplinarity and intersectionality, both environmental justice and sustainability are disciplines and movements that have tended to take a microanalytical—and thus, mechanistic—approach to problem-solving, operating on a neighborhood-by-neighborhood, industry-by-industry, institution-by-institution or project-by-project level. Asking different questions on a both a micro- and macro-level has long been a hallmark of feminist methodology, positioning a critical ecofeminism to make significant contributions to ongoing sustainability discussions that do more than "add and stir" considerations of gender, sexuality, and species. Critical ecofeminism is rooted in a relational standpoint that illuminates inequalities from the personal to the political—ecological, economic, sociopolitical—promoting just and equitable relations by raising questions such as, who benefits, and who pays? Do the means and actions match the professed goals? Who and what is missing from this story (Kheel 1993)? Where does this material come from, and where does it end up—who handles the "waste" (Smith 1997; Nhanenge 2011)? Does this activity promote the flourishing of all those involved, from production to consumption and waste disposal (Cuomo 1998)? And, what model of selfhood lies at the root of this action? Plumwood's critique of "economic man" as a manifestation of western culture's Master Model and the many operations constructing this self-identity—hyperseparation, backgrounding (denied dependency), radical exclusion, incorporation, instrumentalism, homogenization—illuminate and revise the heteronormative and humanist arguments of even a "just sustainability."

At present, mainstream sustainability discourse defines its key terms of *ecology*, *economy*, and *society* from a humanist and mechanistic perspective—if these terms are defined at all. On matters of *ecology* and sustainability, there is little discussion of the intrinsic and ecological values of a climax forest and its multispecies relations, apart from its functions as a carbon sink for offsetting human industrial practices, or its associated instrumental values as a place of beauty, recreation, and source of future pharmaceuticals: that the forest may also be home for indigenous humans practicing subsistence lifestyles is not part of the definition. *Ecology* is defined as distinctly separate from human identity, and discussed as a "resource" or as a set of nonhuman systems to be sustained or depleted based on human "needs." Discussions of *economic* sustainability tend to look at traditional cost/benefit analyses, frequently invoking slogans like "going green makes green too!" and suggesting that business practices are sustainable if they are not merely ecological but also more profitable than business-as-usual. *Society* is seldom differentiated in terms of sexuality, ethnicity, gender, or gendered class; rather, it is used as a mass term that variously refers to stakeholders, investors, consumers, or the community where a business operates. "Sustainability" seldom includes the workers at the business site, or those workers and environments farther removed, who either supply the "sustainable" business with materials or who dispose of the wastes after consumption. In short, without analytic tools from the environmental humanities and critical ecofeminism, sustainability discussions still rely on neoclassical economics and the autonomous individual of liberal political theory.

As feminist economists have argued, "the model of 'economic man' as a separate, autonomous, detached, competitive and primarily self-interested individual is antifeminist, anti-ecological, and oppressive of those who are 'other' than economic man" (McMahon 1997; Nelson 2007; Perkins and Kuiper 2005; Waring 1988). Moreover, it is no accident that "economic man came to maturity in the heyday of colonization" to describe the "white, privileged male whose 'autonomy' was predicated on the oppression of women, nature, and non-white persons, and the destructive colonization of indigenous peoples' lands" (McMahon 167). There are limits to an ecological economics that remains rooted in autonomous individualism; it will have 'trouble with relationships" whether those of ecological justice or social justice, gender justice or interspecies justice. In sum, neoclassical economics "disguises the ways in which the market and economic man are dependent on hidden transfers from nature and unpaid work" (McMahon 172).

Feminists have long observed the ways that western culture defines ecology as separate from culture and humanity, yet locates women, people of color, children, and nondominant others together, as "closer to nature" (Gray 1979; Griffin 1978; Kolodny1975; Warren 1990). The dualisms of Cartesian thought also align gender and economics along the ecology/culture binary, mapping economically visible labor and production in the public sphere, and economically invisible labor, materials, and reproduction in the private sphere. As feminist economists have observed, women, nature, and colonial entities share similar treatment in neoclassical economics; they are backgrounded and treated as a resource for meeting "human" needs: "the bearing and raising of children, and the care of the aged and sick—traditionally women's responsibilities—are, like nature, too unimportant to mention" in national and global accounting systems (156).[23] Perhaps part of the problem is that, until rather recently, white women and children in Western culture have been classified as property of the Master, albeit with greater status than the other "property"—slaves, colonized others, and nature. Fundamental to neoclassical economics is the human relationship to the earth as "property" to be bought, sold, and owned, rather than as a living agent of *vibrant matter* (Bennett 2010)*,* or a "community to which we belong" (Leopold 1946).[24] Plumwood argues that the "Lockean account of the incorporative self upon which capitalism is based" is integral with the Lockean theory of property that embodies assumptions about the "emptiness and nullity of nature itself" and erases "those human others counted as nature" (2002, 214–17). Plumwood's critical ecofeminism envisions a relational self, dynamically interconnected with an agential nature that is far from the inert, lifeless, and mechanistic conception of property—rather it is a nature that actively coconstitutes earthothers. In her materialist spirituality of place, Plumwood describes land ownership as "two-way" practice "in which you belong to the land as the land belongs to you," and where remoteness is dissolved in recognizing "the communicativity and intentionality of more-than-human others" which is "key to the power of place" (230).

In her discussion of "shadow places," Plumwood (2008) brings forward the backgrounded, multiple, and complex network of places that support human lives. She proposes a "critical bioregionalism" that helps "make visible north/south place relationships" and clarifies "relationships of domination metaphorised as place, especially sacrificial and shadow or denied places." As Plumwood explains, the

> dissociation of the affective place (the place of and in mind, attachment and identification, political effectiveness, family history, ancestral place) from the economic place that is such a feature of the global market is yet another manifestation of the mind/body dualism. (Plumwood 2008)

As an alternative and more accurate view, she argues for place as "an active agent in and co-constituter of our lives," and "a process in which the energy of others is actively invested." In one of her last published essays, Plumwood proposed "a place principle of environmental justice":

> an injunction to cherish and care for your places, but without in the process destroying or degrading any other places, where 'other places' includes other human places, but also other species' places. (Plumwood 2008)

As a foundation for ecological citizenship, Plumwood's "place principle" grounds human identities in ecological relationships, reminding us of the moment-to-moment flows of energy and matter required for sustaining all life on earth. Her work radically reconceives sustainability.

REIMAGINING SUSTAINABILITY

Ecofeminists and environmental justice scholars alike have voiced skepticism about sustainability paradigms, raising questions such as *"what* is to be sustained?" and "for *whom?"* and "for how long?" As Sherilyn MacGregor (2009) observes, for the past fifty years both feminism and ecofeminism have proposed *transformative* processes that are about moving "toward an egalitarian and non-oppressive world rather than about keeping things the way they are forever" (2009, 469). Comparing our current ecological situation to the *Titanic* for its "technological hubris and decision-making disaster in the face of risk," Val Plumwood argues that the term "sustainability" has tended to "obscure the seriousness of the situation" with the hope that we could sustain anything without enacting significant system transformation (2002, 1). As she wryly notes,

> The *Titanic* myth is liberal-democratic, maintaining a story of equality of consequences, of elite heroism and self-sacrifice, of millionaires and other men standing back while women and children were saved. But in the real ecological world on which we are passengers, unlike the *Titanic*, the millionaires don't go down with the ship, and it's certainly not women and children [who are saved] first. (Plumwood 2002, 2)

Sustainability discussions tend to imply that "it is in the interests of *humanity in general* to work toward sustainability," when ecofeminists have rightly pointed out that "we are most certainly not all in this together" (MacGregor 2009). In fact, it's possible that some societies, environments, and species would be better off if the high-consumption lifestyles of industrialized western cultures were *not* sustained any longer.

Make no mistake: Reimagining sustainability is critical for our collective survival. But how shall we undertake that visioning process if it relies on excluding the majority of earth's inhabitants, and over half of humanity's collective wisdom heritage?

In place of neoclassical economics, what sustainability—a just ecofeminist sustainability—requires is a *transcorporeal economic accounting* of the ways that our social, economic, and political practices are racialized and gendered, and can be used to either promote the *flourishing* or the languishing of all earthothers. Recognizing the agency and transcorporeality of earthothers, a reinvigorated sustainability will reject concepts of "environment" that reify self-other dualisms, replacing them with relational earth identities. The public/private environmental, economic, and gendered dualisms noted by ecofeminists must be replaced with environmental justice concepts of a continuity of relations among "where we live, work, play and pray"; the focus of "sustainable management" must shift to sustainable *dialogues*, using ecologically, economically, and socially democratic participatory decision-making to enhance listening, awareness, and consideration of transcorporeal eco-socio-economic relations. Just, sustainable, and ecofeminist economics will reject the linear model of neoliberal economics and replace it with the indigenous and ecological model of the circle, where "waste" is no longer a concept, and sustainability is enacted through transformations of repurposing, composting, and reusing former "waste" in new materials beneficial to an ecological community.[25] This new sustainability will prohibit economic "profits" based on theft: whether it is oil from the Arctic and Nigeria, Alberta's tar sands and North Dakota's Bakken fields; the prison labor extracted from an incarcerated population that is disproportionately dominated by Black men,[26] or the sweatshop labor of poor and third world women that results in a century of disasters, from the 1911 Triangle Shirtwaist Factory fire in New York to the 2013 Rana Plaza collapse in Bangladesh; or the theft of self-determination, freedom of movement, the ability to nurture offspring to maturity, and life itself from billions of farmed animals. It will count as loss the formerly lucrative practices of sex trafficking and organ trafficking, the enslavement of animals in zoos and scientific labs and agriculture, and the "free" labor of prisoners in cleaning up tar sands "spills." Defining *society* as referencing a transspecies diversity of citizen identities, this just sustainability will require an inclusive, ecological, economical, and participatory democracy.[27]

Current discussions of sustainability among elite white professionals have the untapped potential to function as a stealth operation, transmitting the small winged seeds of a more critical, just and ecofeminist social transformation. Given the immediacy of climate change, it's high time to open up these conversations.

NOTES

1. According to David Orr, ecological literacy is a transformational approach to sustainability, honoring the "connections between people of all ages, races, nationalities, and generations, and between people and the natural world"; it comprehends "the interrelatedness of life" and "the ways in which people and whole societies have become destructive"; it recognizes "the speed of the crisis that is upon us" and "implies a radical change in the institutions and patterns that we have come to accept as normal" (1992, 93–95). Preceding Orr's work by almost fifty years, Aldo Leopold's (1949) land ethic transforms both the human-environment relation and the human practitioner: "a land ethic changes the role of *Homo sapiens* from conqueror of the land community to plain member and citizen of it. It implies respect for his fellow members, and also respect for the community as such." While revolutionary for its era, Leopold's analysis ignores issues of race, gender, sexuality, species, and nation that are central to a just feminist sustainability; yet the potential for interrogating these issues in relation to a land ethic is latent in his writings, as educators at The Leopold Foundation in Baraboo, WI, have recently developed the land ethic in terms of environmental justice.

2. See Hamline University's "What is Sustainability?" at https://sites.google.com/a/hamline.edu/committee-sustainability/what-is-sustainability. There are at least

282 visual models of sustainability compiled in Samuel Mann, *Sustainability: A Visual Guide* (NewSplash, 2011), available at http://computingforsustainability.com/2009/03/15/visualising-sustainability/. There are no "approved" sustainability diagrams on the AASHE website; I chose the Hamline University diagram because it is both descriptive and simple, unlike many others.

3. See http://www.aashe.org/files/aashe_annualreport2013.pdf .

4. This information is current as of June 2015, available at http://www.aashe.org/resources/academic-programs/.

5. Agyeman's innovative work bridging environmental justice and sustainability perspectives still backgrounds feminist considerations of gender, sexuality, and species, giving these occasional references that are add-ons rather than centralizing these as crucial elements of a just sustainability.

6. See "Annual Report: 2013," Association for the Advancement of Sustainability in Higher Education, accessed at http://www.aashe.org/files/aashe_annualreport2013.pdf on 1/5/2017.

7. See *Friends of the Earth: Tar Sands*, at http://www.foe.org/projects/climate-and-energy/tar-sands, and WorldWatch Institute: Tar Sands Fever, at http://www.worldwatch.org/node/5287.

8. See also the Indigenous Environmental Network on tar sands at http://www.ienearth.org/what-we-do/tar-sands/ and http://www.theglobeandmail.com/news/national/oil-sands-pollutants-affect-first-nations-diets-according-to-study/article19484551/.

9. See also "Tar Sands Solutions Network: Climate Impacts" at http://tarsandssolutions.org/tar-sands/climate-impacts.

10. See also Minnesota House of Representatives Information Brief, *Minnesota's Petroleum Infrastructure: Pipelines, Refineries, Terminals* (June 2013) at http://www.house.leg.state.mn.us/hrd/pubs/petinfra.pdf.

11. See Kayne, Eric. "Defending Fenceline Communities from Oil Refinery Pollution." *Earthjustice* (2014). Case 2180, 3065. Accessed at http://earthjustice.org/our_work/cases/2014/defending-fenceline-communities-from-oil-refinery-pollution# on 6/23/2016.

12. See http://www.democracynow.org/topics/flint_water_crisis for the full report on Flint's water crisis (accessed 6/23/2016).

13. EarthJustice, *Community Impact Report Addendum A: The Toll of Refineries on Fenceline Communities*, October 28, 2014, accessed at http://earthjustice.org/sites/default/files/files/10.28.14%20EPA%20Refinery%20Risk%20Review%2003_Addendum%20A%20-%20Community%20Impact%20Report.pdf on 6/23/2016.

14. See "Environmental Justice," EPA: US Environmental Protection Agency, accessed at https://www.epa.gov/environmentaljustice on 6/23/2016. For a wealth of documentation on the Environmental Protection Agency's failures to provide Civil Rights protection for communities of color, see Vallianatos' (2014), as well as "Environmental Justice / Environmental Racism" at *EJnet.org: Web Resources for Environmental Justice Activists*, accessed at http://www.ejnet.org/ej on 6/23/2016.

15. See "Environmental Justice / Environmental Racism" at http://www.ejnet.org/ej.

16. Four years after Oliei's resignation, in 2010, the state of Minnesota filed a lawsuit against 3M for polluting four aquifers, a 139-mile stretch of the Mississippi River, and over a dozen lakes, contaminating the water supply for more than 125,000 people in the Twin Cities area; by 2014, the lawsuit stalled in legalities and was disqualified (Anderson 2014).

17. The thirty-page legal complaint filed by Minnesota State Attorney Lori Swanson can be found at http://minnesota.publicradio.org/features/2010/12/documents/3m-swanson-lawsuit.pdf

18. For Cottage Grove demographics, see U.S. Census Bureau report, at http://quickfacts.census.gov/qfd/states/27/2713456.html.

19. I am grateful to Corrie Ellis for sending me Molotch's essay.

20. See also the *Canada History Project: Effects of the Fur Trade* at http://www.canadahistoryproject.ca/1500/1500-13-effects-fur-trade.html

21. Plumwood often used the term "anticolonial" rather than "postcolonial," possibly because the latter term was less familiar. I use both terms, preferring anticolonial for the way it reminds us that colonialism is in no way "post"—it continues in the western industrial capitalist treatment of earthothers. In "Androcentrism and Anthropocentrism: Parallels and Politics" (1996), Plumwood uses the five operations of the Master Model to illustrate the linkages operating in the "otherization" of colonialism based on gender, race, class, and indigeneity.

22. For the association of animals with slaves, see the Willie Lynch letter, "The Making of a Slave," delivered in the colony of Virginia in 1712. Lynch uses the slaveowners' familiarity with "breaking" horses to explain how to "break" slaves—starting with the women—to ensure their complete submission.

23. See also Julie A. Nelson (1997). The accounting failures of neoliberal economics in valuing the reproduction of nature and of women are addressed throughout feminist and ecofeminist economics (Waring 1988; Mies and Shiva 1993; Salleh 1997).

24. The insights or new materialists are novel primarily to western culture, as the interconnectedness of all life is integral to most indigenous cultures.

25. See Michael Braungart and William McDonough, *Cradle to Cradle: Remaking the Way We Make Things* (2002) and Janine M. Benyus, *Biomimicry: Innovation Inspired by Nature* (2002) for perspectives that transform our linear economic and production models into a circular economy advocated by many indigenous cultures and environmentalists alike.

26. See Michelle Alexander, *The New Jim Crow: Mass Incarceration in the Age of Colorblindness* (2012), Ava Duvernay's documentary film, "13th" (2016), and Bryan Stevenson, *Just Mercy: A Story of Justice and Redemption* (2015) for the history of institutionalized racism in the United States as enacted through slavery, Jim Crow segregation, and mass incarceration, respectively.

27. Ecofeminists have discussed this tripartite development of citizenship in depth (Plumwood 1993; Gaard 1998; Sandilands 1999).

Chapter 2

Plants and Animals

"Do we need a 'plant ethics' that responds to vegetal instrumentalization in an allied manner to the ways animal ethics has responded to the animal-industrial complex?" (Adamson and Sandilands 2013).[1] At first glance, such a question suggests either an allied plant ethics movement, or a pending omnivore backlash against the decade-long success of animal studies, launched by Jacques Derrida's "The Animal That Therefore I Am" (2002) which catapulted vegan perspectives into academic credibility. Like animal studies, plant studies scholarship has been ongoing for some time, but only recently has emerged as a cutting-edge academic field. One could compare Peter Singer's *Animal Liberation* (1975) with Peter Tompkins and Christopher Bird's *The Secret Life of Plants* (1973) as foundational works for two movements that would later become recognized as companion branches of posthumanist thought, though in thankless academic fashion, the founders of each branch have been strongly critiqued, Singer for his human rights-based moral extensionism, and Tompkins and Bird for their "new age" unscientific speculations. Invoking the field of "critical plant studies" and author of *The Omnivore's Dilemma* (2006), the carnivorous locavore Michael Pollan, Adamson and Sandilands anticipate this comparison in their description of a "Vegetal Ecocriticism" preconference seminar for the Association for the Study of Literature and Environment (ASLE) 2013 biannual conference:

> As critical animal studies and animal rights scholars/activists have effectively worried constitutive boundaries between human beings and other animals, plant studies scholars have questioned the similarly political line between plants and animals: plants communicate, move, decide, transform, and transgress in ways that are sometimes uncomfortably "like" animals (including humans), and sometimes so completely Other to animality that conventional metaphysical principles are radically denaturalized.

The phrase "similarly political line between plants and animals" might raise alarms among animal studies scholars and critical ecofeminists alike, and understandably so. Vegan and vegetarian feminist ecocriticism has a substantial history, starting with Carol Adams' discussion of Frankenstein's vegetarian monster in *The Sexual Politics of Meat* (1990), and continuing through the work of many vegan and vegetarian feminist ecocritics (i.e., Armbruster 1998; Chang 2009; Donovan 1991, 2009; Gaard 2000, 2001, 2013). But Simon Estok's (2009) essay in the flagship journal of ASLE finally threw down the gauntlet. Defining "ecophobia" as an "irrational and groundless hatred of the natural world," Estok argued that "ecophobia is rooted in and dependent on anthropocentric arrogance and speciesism"; thus, it is "difficult to take seriously . . . the ecocritic who theorizes brilliantly on a stomach full of roast beef on rye" (2009; 208, 216–17). Four years after Estok placed the legitimacy of carnivorous ecocritics under scrutiny, the emergence of Randy Laist's *Plants and Literature: Essays in Critical Plant Studies* (2013) confirms that vegan/vegetarian ecocritics will soon need a response to "vegetal ecocriticism," just as animal studies scholars and critical ecofeminists will need to consider the findings and claims of plant studies. So, what is "plant studies," and do its claims of animal/plant similarity seek to delegitimate the very real suffering of other animal species, and place human food choices on a terrain of moral relativism suitable to a carnist[2] culture? Or will this movement invoke new materialisms' arguments for the animacy of all matter, and explore our ethical responsibilities to agential others, as Adamson (2014) and Sandilands (2016) later argued?

Central to these questions are the key terms *meat* and *species*.

PLANT STUDIES: A NEW FIELD EMERGES

Although articles formulating the emergence of plant studies had already begun to appear in journals such as *Quanta* (McGowan 2013), *Mother Earth News* (Angier 2013), *Journal of the Fantastic in the Arts* (Miller 2012), *PAN: Philosophy Activism Nature* (Hall 2012), *Societies* (Gagliano 2013; Ryan 2013), and even *the Journal for Critical Animal Studies* (Houle 2011), plant studies emerged into popular culture through the publication of Michael Pollan's (2013) "The Intelligent Plant" in *The New Yorker*. There, Pollan reports the findings of biologists—molecular, cell, plant—confirming capacities that new materialists call *agency* (Coole and Frost 2010) and suggesting paradigm-shifting parallels to animal capacities as well (see Table 2.1).

Not surprisingly, there are at least two dissenting branches of the field. Anxious to safeguard their work as legitimate science by avoiding animism, conservative (read "hard science") plant scientists interpret their data in very

Table 2.1 Behaviors Confirming Agency in Plants and Animals*

Species/Behaviors Confirming Agency	Plants	Animals
Senses: apparatus and ability to sense and optimally respond to environmental variables	15–20 distinct senses: • sense and respond to chemicals in the air or on their bodies • react differently to various wavelengths of light and shadow • a vine or root "knows" when it encounters a solid object, and vines grow toward supports • plant behavior suggests plants hear the sound of flowing water, and respond to potential threats by generating defense chemicals	5 senses: • Smell and taste • Sight • Touch • Sound
Communication	Plant "signaling" occurs through the release of volatile chemicals, or the production of predator-repelling toxins	Vocalizations, body movements, postures, scents
"Intelligence"	Plants know their environment, location, and other plants nearby Root tips gather and assess data from their environment and respond in ways that benefit the plant community, kin and beyond Store information biologically, through molecular wrapping around chromosomes (epigenetics) "Distributed intelligence" through root networks; know their environment; may use fungal networks to nourish seedlings and even trade nutrients across subspecies	Brain, neurons, nervous system, consciousness Ability to reason, judge Memory & learning: laying down new connections in a network of neurons
Self-identity	Can lose up to 90% of their bodies without being killed	Self-awareness as individuals, family and species members
Arguments for uniqueness (and hence, moral standing, moral consideration, and possibly "rights")	Plant signaling: molecular "vocabulary" releases to signal distress, deter or poison enemies, recruit animals to perform services (i.e. pollination)	Ability to feel emotions, i.e., love, anger, loyalty, joy, playfulness, grief, depression, appreciation of beauty, loneliness, compassion, jealousy, regret, sociality

*Data sources from Angier 2013; Bekoff & Goodall 2007; Bekoff & Pearce 2009; Chamovitz 2012; Gagliano 2013; Pollan 2013; Ryan 2012.

humanist ways that preserve the animal/plant species hierarchy, rejecting the terms "plant communication" and "plant neurobiology" for "plant signaling," and "learning" for "adaptation" (Pollan 2013). More progressive plant stud- ies scholars (read "humanities"), however, suggest not only that we should "stop anthropomorphizing plants" but actually "try instead to think like them, to phytomorphize ourselves" (McGowan 2013). Challenging evolutionary biology's misuse of the concept, "survival of the fittest," Monica Gagliano concludes that even "the very competitive evolutionary process of natural selection involves cooperation," and "cooperation and competition can coex- ist" because among plants, more cooperative, "collective associations are indeed an ecologically common state of affairs" (2013, 153).[3] "Thinking plant-thought shoves us in a better way than thinking animal-thoughts does," argues Karen Houle in the *Journal of Critical Animal Studies*, since the eco- logically "'correct unit' of analysis is not the individual, nor the dyad, but 'the assemblage'" (2011, 111).

Though not explicitly drawing on queer studies, Houle's argument uses a posthumanist methodology that compares favorably with queer methodology (i.e., Browne and Nash 2010). As Cary Wolfe (2009) explains, posthumanism involves both content and method:

> one can engage in a humanist or a posthumanist practice of a discipline. . . . Just because a historian or literary critic devotes attention to the topic or theme of nonhuman animals doesn't mean that a familiar form of humanism isn't being maintained through internal disciplinary practices that rely on a specific schema of the knowing subject and the kind of knowledge he or she can have. So even though your external disciplinarity is posthumanist in taking seriously the exis- tence and ethical stakes of nonhuman beings (in that sense, it questions anthro- pocentrism) your internal disciplinarity may remain humanist to the core. (572)

Ecofeminist, posthumanist, queer, and trans* methodologies alike reject the essentialist, unified Cartesian human for a socially constructed plurality of continually shifting identities and selves,[4] and their methodologies can be seen in Houle's approach to plant studies. For example, Houle provocatively rejects the humanist and heteronormative hegemony of mutualism as a frame- work in plant studies, challenging the conceptual gesture that defines plant behaviors as "communication" only in mutualist dyads: "if the benefits to the emitter and receiver [of plant signals] are not equal and not mutual, the description of the plant behavior is downgraded from 'communication' to 'eavesdropping'" and the "third party is called a 'cheater'" (109). Instead, Houle suggests framing these communications not as "illicit" but as "actions of generosity and gift . . . spontaneous, non-meritocratic . . . uncontainable excess" (109). Pointing to the "permanent and varied role of organic and inorganic thirds and fourths in every communication mechanism" (110),

Houle invokes Deleuze and Guattari's concept of "unholy alliances" to describe plant relations as "a radical collectivity" that transforms sociality and kinship "beyond any simple sense of between" to a broader "among" (111). Houle advocates "becoming-plant" for the ways it "opens up thinking about relations as transient alliances rather than strategies," and "credits the accomplishment of identity and intimacy as *a radically collective achievement*" (112). These arguments fit well with new materialism's concept of *transcorporeality* (Alaimo 2010) as well as queer theory's fluidity of identity, sexuality, and community.

Leading ecocriticism's vegetal branch of "critical plant studies," Catriona Sandilands (2014) conceives of plant studies as emerging from and companioning critical animal studies. Her work on *Queer Ecology* (Mortimer-Sandilands and Erickson 2010) has explored the ubiquitous presence of queer animals that usefully complicates heteronormative assumptions about sexuality, embodiment, and authenticity. Following models of queering animal studies, Sandilands (2014) proposes the concept of "botanical queers" that illuminates how plant lives offer the potential to complicate heteronormative (and humanist) conceptions of identity, kinship, and time. In "Floral Sensations" (2016), Sandilands argues "we cannot consider ethical relations to plants without a willingness to engage with their profoundly different articulations of life, in specific contexts, and involving close attention to the ways in which they act in specific biosocial communities" (235). Sandilands' version of critical plant studies describes human-animal ethics as crafted within an anthropocentric moral extensionism, and argues for the inapplicability of this strategy based on plants' wholly different selfhoods, an approach that leaves human-animal ethics intact. But other approaches may be less benign.

For example, Canadian science writer Elaine Dewar (2013) seems almost gleeful in speculating that the proposal that "plants think" will be "anathema to vegans," and since chemistry professor Susan Murch has called the volatile chemical signals from wounded plants "screams," Dewar warns, humans should "remember that, the next time you rip a carrot out of the garden." It appears that "a specter is haunting animal studies," writes T. S. Miller, "the specter of cellulose" (2012, 460). Miller rightly criticizes the humanist methodology of animal studies, arguing that "it is zoocentrism and not simply anthropocentrism, the bugbear of animal studies," that defines human identity (463). Miller agrees with Matthew Hall's view that "zoocentrism helps to maintain human notions of superiority over the plant kingdom in order that plants may be dominated. It is a crucial dualising force, responsible for depicting plants as inferior beings and as the natural base of a human-dominated hierarchy" (Hall 2011, 6). Rejecting what earlier plant studies scholars (Wandersee and Schussler 1999) rightly term "plant-blindness" in Western culture, Gagliano condemns our current state of "vegetal disregard" for "plants, whose fundamental role is to ensure continuity of life on Earth" (2013, 149). When we "contemplate and confront the vegetal in the human," Miller believes, we will advance the posthumanist project of overturning hierarchies, "strike at the root of humanity's instrumentalist domination of plants . . . [and] recognize kinship with plants [which] will inevitably alter how we think about our use of them" (462).

As with animal studies articulations of posthumanities, plant studies perspectives can be understood through their genealogies. The science studies perspectives in plant studies trace their field from "electrifying discovery" in 1983, through "decisive debunking" in 1984, and on to "resurrection" by 1990 (McGowan 2013). Companioned by more popular science texts such as Michael Pollan's *The Botany of Desire* (2001) and Daniel Chamovitz' *What A Plant Knows: A Field Guide to the Senses* (2012), this science studies branch has gained academic attention through professional organizations such as the International Society of Plant Signaling and Behaviour, whose annual conference in summer 2013 drew scholars from over forty countries, and was discussed in the popular debut of plant studies (Pollan 2013). Prominent scholars such as Susan Murch (Canada Research Chair in Natural Products Chemistry), Catriona Sandilands (Canada Research Chair in Sustainability and Culture), science writer Elaine Dewar, and Basque philosopher Michael Marder each with new plant studies research published or in press, and Marder's new Rodopi Press series on "Critical Plant Studies: Philosophy, Literature, Culture" already producing its first volume on *Plants and Literature* (Laist 2013). Philosopher Michael Marder's *Plant-Thinking: A Philosophy of Vegetal Life* (2013) brings new materialism's concepts of agency and the posthumanities' decentering of the human to develop critical plant studies' redefinition of species as existing on a continuum that is more alike than different, with shared ancestry.

In articles introducing the new field of plant studies, both the purposes for plant studies research and the dominant standpoints seem overwhelmingly humanist and masculinist, instrumentalizing plants for new technologies, or theorizing about plants in ways that benefit humans. For example, Pollan (2013) cites a disproportionate number of male scientists (women number only seven out of twenty-seven researchers cited), and gives "poet-philosopher" Stefano Mancuso of the University of Florence's International Laboratory of Plant Neurobiology greatest prominence in philosophical discussions about how to interpret the science of plant studies. According to Mancuso, the reason to study plant behavior is because "we stand to learn valuable things and develop new technologies," perhaps "design better computers, or robots, or networks," harness plants for computational tasks or send them to other planets for exploration. Masculinist perspectives in plant science lead to questions that are limited and ultimately humanist: for example, observing "interplant communication" through the release of volatile chemicals, ecologists Richard Karban and Martin Heil wonder "why should one plant waste energy clueing in its competitors about a danger?" and conclude that "plant communication is a misnomer" and should be called "plant eavesdropping" or even a "soliloquy" (McGowan 2013).

From a critical ecofeminist perspective, the plant studies genealogy as currently presented is a Euro-patrilineage that parallels the developments and omissions of animal studies (Fraiman 2012; Gaard 2012). Presented as if the field emerged only recently, it erases not only the methodology and findings of Nobel prize-winner Barbara McClintock (Keller 1983) and marine biologist Rachel Carson (1951, 1962), but also two centuries of work on human-plant relations explored by women gardeners, scientific illustrators, animal writers, and ecological artists (Norwood 1993; Norwood & Monk 1987; Anderson 1991; Anderson & Edwards 2002; Gates 1998). Even more significantly, elite plant studies genealogies largely omit indigenous non-Western perspectives—contemporary ecoactivists and writers such as Winona LaDuke, Tom Goldtooth, Gloria Anzaldúa, Chico Mendez, Ken Saro-Wiwa, and many others—whose cultures never made the Aristotelian divisions of humans from the rest of life,[5] and thus whose writing about plants, animals, and ecology has never needed material feminism's recuperative concepts of "transcorporeality" (Alaimo 2010) or "naturecultures" (Haraway 2003) to illuminate *All Our Relations* (LaDuke 1999). Those plant studies scholars who do attend to indigenous perspectives currently work in the humanities wing of plant studies—anthropology (Kohn 2013; Viveiros de Castro 2004), ecocriticism (Adamson 2014), philosophy (Hall 2011; Marder 2013), gender, and cultural studies (Plumwood 2000, 2003, 2012; Sandilands 2014, 2016).

As usual in environmental studies, the findings of the environmental sciences wing are crucial yet insufficient, and may in fact be operating on distorted premises, purposes, or hypotheses; they lack the contextual, philosophical, and political reflectiveness of the environmental humanities. Bringing forward feminist animal studies in dialogue with Val Plumwood's critical ecofeminist work on indigeneity and vegetarianism, Hall's philosophical botany, and the queer/posthumanist/feminist approaches of Hall, Gagliano, and Sandilands is needed in cultivating a critical ecofeminist perspective on human, animal, plant, and ecological relations. Rethinking these relations augments our understanding of "species" and "meat."

DEFINING "SPECIES" AND "MEAT"

In animal studies, the key terms "species" and "meat" reference a wealth of field-defining and movement-building conceptual groundwork. Peter Singer's

(1975) key concept of *speciesism* defined as "a prejudice or attitude of bias toward the interests of members of one's own species and against those of members of other species" (7) has been used to build analogies to other humanist structures of oppression—racism, sexism, classism, ethnocentrism, anthropocentrism—and to persuade progressive communities of feminists, civil libertarians, labor activists, antiracist allies, and environmentalists alike to recognize and reject mutually reinforcing forms of hierarchical and dominative thinking, or what ecofeminist philosopher Karen Warren first called the *logic of domination* (1990).[6] The year 1990 was equally significant for feminist animal studies: the leading journal of feminist scholarship, *Signs*, published Josephine Donovan's essay describing more than a century of vegetarian women's activism and theory connecting a feminist ethics of care with animal defense, and Carol Adams' *The Sexual Politics of Meat: A Feminist-Vegetarian Critical Theory* explored the ways gender dualisms linking men, meat eating, and virility in opposition to women, vegetables, and passivity reinforce the subordination of women and nonhuman animals through processes of objectification, fragmentation, and consumption. Adams' concept of animals as the *absent referent* of meat eating, and female animals as the *absent referent* in not only meat eating but also dairy and egg production gave vegan feminist language to what it means to become "a piece of meat." Shortly afterward, *Ecofeminism: Women, Animals, Nature* (Gaard 1993) became the first volume to link feminist animal studies with ecofeminism by placing species at the center of ecofeminist praxis, and launching a decade of debate among feminists and ecofeminists about the place of animals in an antiracist, nonessentialist and postcolonial ecofeminism (Gaard 2003, 2011).

In 2014, feminist animal studies still differs from the newer, yet mainstream branch of animal studies, and is more closely allied with critical animal studies. Whereas animal studies has tended to investigate human-animal relations from an academic perspective, both feminism and critical animal studies are movements for justice; many critical animal studies scholar-activists are also feminists. Unlike animal studies, at the heart of feminism is the centrality of praxis, the necessary linkage of intellectual, political, and activist work. Feminist methodology has challenged the male bias masquerading as objectivity in science, and worked to undermine the fit of science with dominant modes of exploitation and oppression that use science to benefit elite humans, often at the expense of disenfranchised humans and experimented-upon animals (Harding 1987, 1991; Keller & Longino 1996; Stanley 1990). In stark contrast, feminist methodology requires that feminist research puts the lives of the oppressed at the center of the research question, and undertakes studies, gathers data, and interrogates material contexts with the primary aim of improving the lives and the material conditions of the oppressed. When feminists attend to "the question of the animal," we do so from a standpoint that centers other animal species, makes connections among diverse forms of oppression, and seeks to put an end to animal suffering—in other words, to benefit the subject of the research (Birke & Hubbard 1995).

On the surface, contemporary plant studies may share a commitment to plant well-being as well. Its scholarship challenges the definitions of "species" and "meat," charging that animal studies scholarship (across the branches of animal studies) has only "moved the line" of moral considerability, performing a humanist moral extensionism that includes other animal species but places plants outside the bounds of moral consideration—effectively, *treating plants like meat*. In animal studies, plants are "backgrounded," writes Houle (91–92), and Matthew Hall (2011) agrees, invoking Val Plumwood's (1993) theory of the Master Model construction of both the Master identity and a logic of domination (Warren 1990) that operates through Plumwood's five linking postulates of the Other's homogenization, hyperseparation, backgrounding, instrumentalism, and denied dependency. In short, animal studies may be liberating for animals, but oppressive for plants, thus perpetuating humanism. Not only are plants "kin" to animals, but the presence of carnivorous plants, herbivorous animals, sea anemones, algae, and fungi (Tsing 2012) which ambiguously display attributes of both plants and animals have all begun to blur the demarcation of "species" dividing plants from animals, first asserted by Charles Darwin's theory of common descent in his *On the Origin of the Species* (1859). This radical continuity, transcorporeality, and kinship across plant and animal species gives rise to several questions: First, as feminist methodology suggests, we can ask, do plants benefit from plant studies? And second, inspired by a queer feminist and posthumanist animal studies, we must ask, without creating a moral underclass, how do we make ethical food choices in light of the fact that all potential "foods" are sentient beings? And finally, in what ways can a feminist methodology be used to study and improve conditions and interspecies relations among plants, humans, and animals, augmenting critical animal studies by responding to the findings and implications of plant studies scholarship?

In considering these questions, there are arguments critical animal studies and ecofeminist scholars would want to endorse, and others to avoid. For example, plant neurobiologist Stefano Mancuso believes that "because plants are sensitive and intelligent beings, we are obliged to treat them with some degree of respect," which means "protecting their habitats" and "avoiding practices such as genetic manipulation, growing plants in monocultures, and training them in bonsai" (Pollan 2013). This standpoint seems consonant with feminist animal studies arguments and feminist methodology. But this standpoint does not prevent humans from eating plants since "*plants evolved to be eaten*," Mancuso asserts, given plants' lack of irreplaceable organs and modular structure. This argument is reminiscent of parallel justifications of human predation on other species (animals) and their *telos* (i.e., they are "meant to be eaten"), whether based on biology (theirs or ours), culture, or need.

Reviewing the arguments for veganism, we find some of these apply also to plants.

1. *They don't want to be eaten.* Countless texts in animal studies confirm that animals don't want to be eaten: their behavior speaks their desire as they run away from hunters, fight against other predators, and struggle to free themselves from zoos, leg-hold traps, science experiments, and other forms of confinement (Hribal 2011). While certain parts of plants evolved to be eaten by animals to aid in plant reproduction—and attract those animals through scent and color—specific plant behaviors suggest that plants don't want their other parts to be eaten. They give off chemical signals when attacked by insects, alerting other plants, and sometimes invoking predatory insects to feed on the attackers; plants also produce toxins altering a leaf's flavor or texture, making it less palatable and less digestible to herbivores (Angier 2013; Pollan 2013). In queer theory and in feminist theory alike, a primary consideration is *consent*: that is, if all parties don't consent to a specific behavior or relationship, this lack of consent signals potential exploitation, oppression, or otherwise ethically dubious relations at work.
2. *They feel pain.* Animals suffer and feel pain, and thus deserve not to suffer: Singer's utilitarian argument has powered animal rights for decades. Animals are also clearly subjects-of-a-life, as Tom Regan has argued: animals feel pain, emotions, and have a sense of selfhood that affirms their intrinsic value and gives them moral rights; they are not to be used as means to an end for others (Regan 1983). Plant studies confirms that plants give off volatile chemical signals when attacked, suggesting plants may experience vegetal versions of fear and pain. Although plants lack a brain and nervous system, their documented behaviors suggest a level of plant intelligence and plant communications that is presently beyond our knowledge. Lack of information does not ethically justify behavior (though it may explain behavior) and does not provide foundation for causing fear and pain without consent.
3. *They have consciousness.* While the consciousness of most animal species involves a sense of selfhood that is simultaneously individual, familial, species-based and relational, plant studies research suggests that plants have consciousness too in that they change behaviors based on environmental information, communicate (or "signal") to other plants and insects, share nutrients, nurture their offspring.

Adapting to plant studies and the questions of human diet, the four frequently invoked bases for veganism—environmental, human health, world hunger, animal suffering—and placing these in conjunction with Deane Curtin's (1991) contextual moral vegetarianism[7] offer some clarity. As Curtin succinctly argues, "the reasons for moral vegetarianism may differ by locale, by gender, as well as by class"; as a "contextual moral vegetarian," Curtin "cannot refer to an absolute moral rule that prohibits meat-eating under all circumstances" (69). For "the point of a contextualist ethic is that one need not treat all interests equally as if one had no relationship to any of the parties" (70), so Curtin can envision extreme situations of starvation or danger where killing another animal (of any species, including human) would be ethical. Curtin also cites environments that do not support a vegan or vegetarian diet as contexts where food ethics will be based on other considerations of relationships, both animal and ecological; he discusses cultures that pay respect to animals unavoidably killed during human agricultural practices (Japanese Shintoism), and cultures that give human bodies back to other animals as food, recognizing our place in the food chain (Tibetans). Instead of using other cultures' food practices as an excuse for Western predation, Curtin emphasizes, "if there is any context . . . in which moral vegetarianism is completely compelling as an expression of an ecological ethic of care, it is for economically well-off persons in technologically advanced countries" (70). To support this view, he cites the impact of first world agriculture and consumption practices on first world and global environments, the role animal agriculture plays in exacerbating human hunger, the exploitation of female animals in the egg and dairy industries, and the fact that first world consumers have a variety of food options that do not perpetuate suffering. It is a moral and political injunction to "eliminate needless suffering wherever possible," Curtin argues (70). His argument has been further supported by the Food and Agricultural Organization of the United Nation's report, *Livestock's Long Shadow* (Steinfeld et al. 2006), and by research at the Oxford Martin School (Springmann et al. 2016), both documenting the deleterious ecological and climate-changing effects of industrial animal agriculture, and the ecological alternative: widespread adoption of a plant-based diet could drop climate change emissions by 63% and make people healthier, too (Harvey 2016). These findings have been popularized and presented in documentaries such as "Meat the Truth" (2008), "Beef Finland" (2012) and "Cowspiracy" (2014). Ample evidence document the fact that eating such animals involves eating the planet, since the production of animals for food requires vast amounts of plants, water, soil, and other animals (Pimentel & Pimentel 2003); it requires the exploitation of low-waged workers in horrific working conditions (Schlosser 2001) and has devastating impact on human health (Robbins 1987; Campbell and Campbell 2006). Moreover, environmental justice theorists have also addressed the ways that speciesism and racism are mutually reinforcing, developing environmental justice ethics that attend to transspecies ethics as central to analysis (Pellow 2016). Ethical eating is not merely a question of eater and eaten, but a question of eater-eaten-environment—and environments are simultaneously ecological, sociocultural, and economic.

Curtin kindly affirms that there may be no moral *destination*, but rather a moral *direction* we can move in making decisions around what counts as food; I call this insight "kind" because it resists the judgmental attitude omnivores ascribe to vegans, and makes visible the fact that all food production involves some death. As Lori Gruen (2014) elaborates,

> We can't live and avoid killing; this is something I think has been underexplored in vegan literature. . . . We harm others (humans and nonhumans) in all aspects of food production. Many are displaced when land is converted for agricultural purposes, including highly endangered animals like orangutans who are coming close to extinction as a result of the destructive practices used to produce palm oil, a ubiquitous ingredient found in a large number of prepared 'vegan' food products.[8] (132–133)

Contextual moral veganism moves first world consumers in a moral direction but does not eliminate the deaths and consumption of some sentient others, both animal and plant, who are more kin to humans than westerners have recognized. An important component of critical ecofeminism, contextual moral veganism is capable of acknowledging the sentience of plants and other ecological beings, and in diverse contexts, placing humans in the food chain as both eater and eaten, pointing to context-specific moral directions that strive to produce the least suffering and greatest care for all involved: humans (industrial, rural, agricultural, indigenous), animals, plants, ecological entities.

Though ecofeminism has conceived of humans as always embedded within specific and diverse environments, developing theory in consideration of both social justice and environmental concerns, contemporary plant studies and new materialism's concepts of agency and transcorporeality raise more specific questions about the place of plants in environmental and dietary ethics alike. A paradigm-fracturing shift is needed here, one that acknowledges human interidentity, inextricable from and supported by a web of relations with sentient, intelligent kin across species. Val Plumwood argued for such a shift in her theory of "ecological animalism" (2012) or "animist materialism" (2009) or what she variously calls a "critical feminist-socialist ecology" (2000, 285) and "critical ecological feminism" (2000, 289). The term "critical ecofeminism" both acknowledges Plumwood and advances upon more recent work bridging animal and plant studies, feminism and ecology, first world and indigenous perspectives that is the prescient hallmark of Plumwood's thinking. A strong antiracist ally committed to challenging white privilege, Plumwood misread the vegan feminism of Adams and Kheel as devaluing the ethics and worldview of indigenous Australians, whose diet (like all indigenous groups) has developed in relationship to their immediate environments, and includes a range of plants and animals, all of whom they regard as kin. In her theory of "ecological animalism," Plumwood (2003) advocated a "context-sensitive semi-vegetarian position" and opposed factory farming but not a subsistence, need-based killing of other animals; the goal of her theory was "situating human life in ecological terms, and situating non-human life in ethical terms." Many aspects of Plumwood's theory are helpful in developing a critical ecofeminism that is responsive to agency and transcorporeality across plant and animal species.

As if anticipating critiques of the humanism in Ursula Heise's (2008) "eco-cosmopolitanism," Plumwood (2003) argues against the "'biosphere person' [whether vegan or omnivore] who draws on the whole planet for nutritional needs defined in the context of consumer choices in the global market" and whose lifestyle is "destructive and ecologically unaccountable." Instead, Plumwood develops a "critical bioregionalism" (2008) that makes visible "the shadow places of the consumer self," those "places that take our pollution and dangerous waste, exhaust their fertility or destroy their indigenous or nonhuman populations in producing our food." In delightful anticipation of queering ecofeminism and plant studies, Plumwood proposes "envisioning a less monogamous ideal and a more multiple relationship to place" which she phrases as an "accountability requirement" involving "an injunction to cherish and care for your places, but without in the process destroying or degrading any other places, where 'other places' includes other human places, but also other species' places." Urging that we "try to see creativity and agency in the other-than-human world around us" (2003), Plumwood simultaneously maintains that "all embodied beings are food and more than food" and yet "no being, human or nonhuman, should be ontologised reductively as meat" (2009). In these statements, Plumwood develops a contextualized ethic of human interidentity as embedded with ecological others, accepting our place in the food chain as both eater and eaten.[9] Considering ourselves as potential prey for other animal species, both in the midst of our lives and after our death, is wholly consonant with Plumwood's critical ecofeminism (1995, 2012).

Westerners are troubled by indigenous views of nonhuman animals, plants, and ecological beings such as rocks, water, and soil as not only sentient, but kin to humans; seen as animism, these views suggest a perspective belonging to the disciplines of anthropology and comparative religions rather than environmental science. Yet, new developments in plant studies and elemental ecocriticism (Adamson 2014; Cohen 2010) recognize a similar animacy and agency in ecological others. In ecoanthropology, scholars such as Eduardo Viveiros de Castro (2004) and Eduardo Kohn (2013) describe the ecological worldview and linked self-identity of indigenous Amazonian cultures, which emphasize nondifferentiation between humans and animals, intercommunicability, and a state of being wherein self and other interpenetrate. As Viveiros de Castro explains, "cultivated plants may be conceived as blood relatives of the women who tend them, game animals may be approached by hunters as affines, shamans may relate to animal and plant spirits as associates or enemies" (466). In such worldviews, humans and animals are interchangeable, becoming one another as a result of death and consumption. For westerners, such eating of one's kin seems like cannibalism, and indeed some vegan and vegetarian activists consider eating other animal species a form of cannibalism.[10] But if plants are also sentient and kin, then all eating becomes the eating of relatives, a significant ethical conundrum for westerners.

Horror movies "reflect a deep-seated dread of becoming food for other forms of life," Plumwood argues, but "as eaters of others who can never ourselves be eaten in turn by them or even conceive ourselves in edible terms, we take, but do not give" (2003). Rejecting this antiecological human supremacy, Plumwood proposes that "in a good human life we must gain our food in such a way as to acknowledge our kinship with those whom we make our food, which does not forget the more than food that every one of us is, and which positions us reciprocally as food for others" (2003). But this reciprocity to the ecological network or system does not reciprocate to the individual being consumed, as Ralph Acampora (2014) argues. Taking Analía Villagra's work on "Cannibalism, Consumption, and Kinship in Animal Studies" (2011) seriously, Acampora explores her proposition that indigenous worldviews "allow for the consumption of fellow animals not in the absence of or in spite of bonds of kinship, but rather because of them" and thus Villagra argues for "becoming cannibal" (50, 52). But as Acampora observes, Villagra's indigenous cannibalism applies to the extended kinship of other species, and does not commend anthropophagy (cannibalism of humans by humans). Whereas Villagra's use of indigenous worldviews leads her to a moral relativism in her concluding discussion of "my delicious pet," Acampora emphasizes the fact that the indigenous worldviews Villagra discusses have clear understanding of kinship distances. It appears there is both a moral direction and moral destination in these contextual and relational practices of eating, living together, and sharing souls. We can honor all our relations and still move in a moral direction that reduces suffering across species and bioregions, though our lives will never reach a moral destination of universal nonharming.

BEYOND MEAT, BEYOND SPECIES

What shall we eat? Eating plants ensures humans *consume less*—fewer plants, and fewer animals—and ensures we cause less suffering for plants, animals, ecosystems, and other humans as well. It ensures we free up more land for all of life—for plants, for animals, for humans to eat and thrive. And, it monkeywrenches climate change. Bringing into dialogue the critical animal studies and plant studies branches of posthumanism through a lens that recuperates the feminist lineage of these branches, critical ecofeminism has much to offer.

It argues that undoing the grasp and hegemony of a carnist culture requires shifting from denial to attentive listening, from alienation to empathy, from capitalist production time to seasonal time, from a heteronormative universalism to a queer multiversalism. It involves a refiguration of selfhood from rationalist individualism to material transcorporeality, releasing the internalized capitalism of a self-worth based on ceaseless production, and replacing it with a selfhood attuned and intraactive with the cycles of seasonal growth and decay; it offers an acceptance of our death as part of these cycles, and locates our place in the food chain as both eaters and eaten, invoking a contextual moral veganism that values individual lives as well as ecosystem relations, and makes food choices that move in directions promoting sustainability and reducing suffering for eater-eaten-ecosystem.

Philosophers, ecoanthropologists, ecocritics, and other scholars of Western culture are using the tools we have inherited as ways to theorize our understanding of other beings, of nonhuman animals, plants, minerals, and other planetary entities. But as Audre Lorde has written, "the master's tools will never dismantle the master's house" (1984). What's needed is a conceptual shift, a "re-think" as Plumwood proposed (2009), resituating our perspectives so that they companion the standpoints of non-Western cultures, other animals, and plants. This shift involves the practice of attentive listening advocated by feminist animal studies scholars (Donovan 1990, 1998) and plant studies scholars as well: it involves learning a different language of embodiment, behavior, scent, and intraactivity. As strategies for seeing "creativity and agency in the other-than-human world around us" (Plumwood 2009), westerners can stand with the struggles and lives of animals, plants, and indigenous peoples; we can utilize tools from the Buddhist practice of mindfulness and the nontheistic principles of the dharma, originating in India but freely offered across cultures.

From Plato and Descartes, westerners have learned to "treat consciousness rather than embodiment as the basis of human identity" (Plumwood 2004, 46). But these elements of being are inseparable, intraactive elements of agency, interconnected with other flows of planetary life. Just as animal studies loosened the grip of humanism for westerners, now plant studies—along with vegetal and elemental ecocriticism—offer additional tools. For example, Michael Marder's *Plant-Thinking* (2013) proposes that we examine the world of plants from their own perspective, a practice of attentiveness that uncovers the meaning of plant life, made evident through the seasonal changes, growth, and cyclical character of being. To cultivate this attentive stance, Plumwood (2005) advocates the practice of decolonizing gardening, "a healthful pursuit that brings gardeners into contact and collaboration with nature and sensitizes them to the earth, the rhythms of the seasons, growth processes and the life and death cycles of living things." When Marder discusses the freedom of plants, he argues that plants' indifference, lack of individualized selfhood, and lack of concern for their own self-preservation make them capable of the freedom of play; thus, attentiveness to plants leads not only to plant emancipation, but our own alongside, enabling us to recognize the rich diversity of perspectives possible when consciousness and thought are understood as a creative and inventive "thinking before thinking" (154). Plants' lack of *individualized* selfhood echoes the Buddhist no-self (*anatta*), a concept that rejects the ideas of a separate autonomous self in favor of an *interidentity*, and is linked to awareness of impermanence (*anicca*) and dependent origination (*paṭiccasamuppāda*), the understanding that all beings and events coarise and pass away. Plants' alleged lack of concern for self-preservation (not entirely accurate if we consider the volatile scents released when plants perceive an attack) shows up in the Buddhist concept of nonclinging, since clinging (*tanha*) is the basis of suffering (*dukkha*).

From a Buddhist perspective, the perceived indifference of plants, like the imputed indifference of rocks and other minerals, might be better understood as *equanimity*, the ability to be present to life without placing conditions on how life shows up. In ecofeminist theory, Karen Warren's (1990) work demonstrates how the logic of domination has been used to rationalize human dominance over plants and rocks (128–129), a dominance she refutes through a rock-climbing narrative that offers a westerner's practice of attentiveness to mineral life (cf. Cohen 2010; Gaard 2007):

> I closed my eyes and began to feel the rock with my hands—the cracks and crannies, the raised lichen and mosses, the almost imperceptible nubs that might provide a resting place for my fingers and toes when I began to climb. At that moment I was bathed in serenity. . . . I felt an overwhelming sense of gratitude for what [the rock] offered me—a chance to know myself and the rock differently, to appreciate unforeseen miracles like the tiny flowers growing in the even tinier cracks in the rock's surface, and to come to know a sense of being in relationship with the natural environment. It felt as if the rock and I were silent conversational partners in a longstanding friendship. (Warren 1990, 134)

Here, Warren's ecofeminist ethics rests not on the selfhood or relative value of the other, but on one's own quality of attention, care, and relationship. An ecofeminist transspecies ethics—addressing animals, plants, minerals, and other planetary life—is a relational ethic, not based on the individualist self-identity or rights of various others.

The emergence of plant studies illuminates the vegetal considerations already present in indigenous cultures, in Buddhism, and in critical ecofeminism, a Western perspective that brings together social and environmental justice, climate justice, and interspecies justice. Ecofeminism's contextual moral veganism offers a useful strategy for making decisions about ethical eating for humans, plants, and animals; its contextual aspect is not a form of moral relativism, nor is it a universal rule. A critical ecofeminism encourages a shift in our thinking and in our being, from a humanist perspective of dominance to an awareness and participation in relations of mutuality and reciprocity that resituate humans in the cycles of planetary life.

NOTES

1. The question appeared in a call for participants attending a preconference seminar on "Vegetal Ecocriticism" held at the ASLE 2013 conference. Since 2013, the seminar coleaders have continued to develop theories on this topic, taking it in quite useful directions, and I am especially grateful to both Joni Adamson and Cate Sandilands for the several emails we exchanged to clarify their different viewpoints with updated essays; our standpoints differ, and I strive here to present theirs accurately (though briefly). Building on her work in indigenous studies, Joni Adamson argues for a wholly different cultural paradigm, a "'cosmo*politics*' capable of mediating dramatically differing (intra- and interspecies) perspectives" (2014, 264). Working within a Western-cultural paradigm, seminar coleader Cate Sandilands argues that to develop "ethical relations to plants," it will be "clearly inadequate to anthropomorphize their capacities and interests in the service of an abstract conception of life or justice" (236). While both ecocritics have produced scholarship advancing a "vegetal ecocriticism," neither scholar has (yet) articulated an environmentally just ethics and praxis for both human-social and transspecies relations in Western industrialized cultures. As this chapter explains, critical ecofeminists are particularly interested in transspecies ecological ethics crafted with attention to the intersections of racism, classism, speciesism, and sexism in Western culture (cf. Pellow 2016).

2. Melanie Joy (2010) uses *carnism* to describe cultures that make meat eating seem "normal, natural, and necessary"—effectively, hegemonic—through a conceptual schema that uses "objectification, deindividualization, and dichotomization" (96, 117). Joy's work popularizes concepts from more sophisticated philosophical arguments such as Val Plumwood's (1993) Master Model and the wealth of groundwork provided by vegan and vegetarian scholar-activists.

3. This book went to press at the time two new texts appeared, *The Green Thread: Dialogues with the Vegetal World* (Vieira, Gagliano and Ryan 2016), an interdisciplinary volume of ecocriticism that develops work on plant agency and biopolitics, and *Plant Theory: Biopower and Vegetable Life* (Nealon 2016), a text that critiques how

animal studies has tended to shift the dualism of culture/nature so that animal species stand with humans, but leaves the nature-others category intact.

4. As queer studies scholars have argued, queer perspectives may endorse but are more likely to differ from the liberal assimilationist goals of the lesbian/gay/bisexual/ transgendered (LGBT) movement for inclusion in heteronormative institutions (i.e., equal marriage, equal inclusion in the military, LGBT human rights legislation, corporate sponsorships for pride). Present in groups such as the Radical Faeries of the 1960s and queer activist groups of the 1990s, ACT UP and QUEER NATION, some queers have expressed resistance to heteronormative assimilation, choosing instead to celebrate queer culture, eschew essentialist dualisms of gender and sexuality, and affirm queer diversities across race, gender, and class (Jagose 1996; Gleig 2012). These perspectives form the base for queer methodology (Browne and Nash 2010). Special issues of the *Journal for Critical Animal Studies* (Grubbs 2012), *Gay & Lesbian Quarterly* (Chen & Luciano, "Queer Inhumanisms," 2012), and *Transgender Studies Quarterly* (Hayward & Weinstein, "Tranimalities," 2015) have recently devoted attention to the sometimes-overlapping intersections of queer theory and antispeciesist praxis (JCAS), queer theory and posthumanisms (GLQ), or transgender theory and animal studies (TSQ). Indeed, transgendered people, captive Africans, indigenous North Americans, African ("exotic") animals and carnivorous plants have long populated the zoos, circuses, and/or freak shows of colonial empires, and merited the dehumanizing and ethically stripped pronoun "it." Eva Hayward and Jami Weinstein in *Transgender Studies Quarterly* (2015) propose "trans*" and specifically "tranimalities" as terms that utilize humanism's exclusion of transgendered and more-than-human animals from consideration, and "enmesh trans* and animals in a generative (if also corrosive) tension leading to alternate ways of envisioning futures of embodiment, aesthetics, biopolitics, climates, and ethics" (201). At this early stage of articulation, it's unclear whether these "tranimalities" will ally themselves with critical animal studies in rejecting industrial animal food factories or carnist cultures that legitimate human-animal exploitations, often through wordplay (i.e., more-than-human animals are not "killable," yet first world omnivore diets remain unchanged).

5. This critique fits well with Joni Adamson's argument for an indigenous cosmopolitics (2014). My view addresses primarily Euro-western plant studies, and argues for bringing the insights of indigenous and non-western perspectives into dialogue with this branch of theory as well. Culturally contextualized ethics are crucial to an antiracist and anticolonial ecofeminism (cf. Gaard 2001).

6. Warren did not include speciesism in that logic for another decade; see Warren 2000.

7. In distinguishing uses of "vegetarian" and "vegan" in the 1990s, I quote Richard Twine's (2014, 206n4) excellent explanation: "During the 1990s . . . the term 'vegetarianism' was mostly used instead of 'veganism.' I would contend three reasons for this. First, I expect some North American writers used vegetarianism but meant veganism. Secondly, since the 1990s . . . vegetarianism has lost a lot of credibility as a consistent ethical position within the animal advocacy movement but at *that* time it was still deemed credible. Thirdly, and relatedly, during the first decade of the twenty-first century, notably in Western countries, there has been an ethical shift toward, and

cultural normalization of, veganism as the preferred and more consistent practice of animal advocates. So much so, that an ecofeminist arguing today for ovo-lacto vegetarianism would suffer from a credibility problem."

8. Unsustainable palm oil requires deforestation and the deaths of orangutans alike, and a strong resistance is organized through The Orangutan Project (http://www.orangutan.org.au/index.htm). Thanks to Kate Rigby for alerting me to this important intersectional activism.

9. The new materialisms have uncovered the various ways that our own bodies are colonized by microbial life, and how illnesses soon to be exacerbated by climate change are a manifestation of other microbial species feeding on humans. The hegemony of humanism prevents nonindigenous humans from conceiving of ourselves as prey, as food, for the duration of our lives, and only after death do some cultures (i.e., Tibetan) offer human bodies as food for other animals. The "eater and eaten" phrase in the text is thus still unequal: over the course of a lifetime, humans do far more eating than being eaten.

10. These are private conversations; to my knowledge, no one has yet theorized this connection besides Ralph Acampora (2014). I noticed this association when indexing my first book (Gaard 1993), and discovering the term "cannibalism" cropped up enough to be indexed; at the time, the prospect seemed too far-fetched to mention.

Part II

ILLUMINATIONS

Chapter 3

Milk

Overcome with diarrhea and intestinal cramps, villagers in Columbia and Guatemala conclude that the powdered milk rations donated by the United States must not be food, and use the powder, mixed with less water, to white-wash their huts (McCracken 1971; Kretchner 1972). In India's Kerala province, a dairy farmer stares with dismay at the huge Swiss Brown cow that has replaced her native dwarf Vechur cow and exponentially increased her costs for feed and veterinary bills (Sainath 2012). Living along the industrialized St. Lawrence seaway where General Motors has been dumping PCBs and heavy metals for over 25 years, Akwesasne midwife Katsi Cook starts the Mothers' Milk Project after discovering alarming levels of persistent organic pollutants (POPs), DDT, and flame retardants in Mohawk mothers' breast milk and in the body-fat of Beluga whales (LaDuke 1999). Fifteen years later, Sandra Steingraber passes a glass of her own breast milk among the delegates at a UN panel hearing on the reproductive health effects of POPs, emphasizing the bioaccumulation of toxins at the top of the food chain, in the bodies of nursing infants (Steingraber 2001). In 1994, milk produced with recombinant Bovine Growth Hormone (rBGH)—Monsanto's Posilac—appears in U.S. grocery stores, approved by the U.S. government's Food and Drug Administration (FDA) and inspiring protests among small farmers, consumer groups, environmentalists, and animal advocates alike (Gaard 1994).

What critical framework is sufficiently inclusive to describe these uses of milk across nations, genders, races, species and environments? One scholarly field well positioned to address milk, food studies argues that food history is a history of ideas, and milk—a commodity the American dairy industry has marketed as "natural" and "wholesome"—is not a homogenous entity, but one that has various meanings and compositions in different historical and cultural contexts. Both Deborah Valenze's *Milk: A Local and Global*

49

History (2011) and Anne Mendelson's *Milk: The Surprising Story of Milk Through the Ages* (2008) trace Western food history over the last 5,000 years, arguing that milk has been crucial to the survival of many Eurasian cultures, whose people could digest animal-derived lactose from cows, sheep, and goats. Religious histories from Hinduism to Catholicism show the importance of milk in spiritual practices past and present: In India, milk is still used to feed the elephant-headed god of wisdom, Ganesha; in Medieval Europe, St. Bernard had a vision of being miraculously fed by the Virgin Mary's breast milk.

Food histories of Great Britain (Atkins 2010) and the United States (DuPuis 2002) in the nineteenth and twentieth centuries confirm that from 1850 onward, milk was commodified on a large scale, with its highly perishable liquid form inspiring urban dairy production, railway transport, and finally doorstep delivery. Until its sterilization or pasteurization in the 1920s, milk was one of the major public health issues of the late nineteenth and early twentieth centuries, transmitting infections of various types, along with epidemic diseases such as scarlet fever, typhoid, and tuberculosis. Milk-fed infants also faced higher morbidity and mortality rates, yet cows' milk continued to be used in orphanages (where no mothers or wet-nurses were available) and in families alike. In her chapter asking "Why Not Mother?" Melanie DuPuis argues that "the rise of urbanization and the rise in artificial feeding of children went hand in hand" (46). U.S. women chose not to breastfeed for reasons that differed across class: middle- and upper-class women were allegedly fragile, with "nerves" that would be disturbed by breastfeeding; additionally, breastfeeding took time away from their social obligations, and combined with pressure from husbands who wanted their wives' attentions, but were otherwise barred by social norms that proscribed sex while nursing. Both very poor and very wealthy women faced another barrier to breastfeeding: inadequate food intake, a result of either poor diet or cultural norms for "dainty" eating. Most urban well-to-do women turned to formula, cows' milk, or wet-nurses. Working-class women could not afford such luxuries, and breastfed their infants unless prevented by economic circumstances.

The pervasive availability of cows' milk today—from grocery stores to gas stations—is a historically unprecedented product of industrialization, urbanization, culture and economics. Without human intervention, fresh cows' milk is largely unavailable for more than part of a year (March to November): cows require nine months for gestation, along with ample pasture and feed in order to produce milk (DuPuis 28–29). Its availability is part of Western industrialized culture's continuing "triumph over nature"; as Atkins concludes, in Britain, milk was "representative of efforts to redraw the boundaries between nature and society" (xix). Far from being "the perfect food," milk offers a narrative about progress and perfection that embodies "the politics of American identity over the last 150 years" argues DuPuis (8), linking "the perfect whiteness of this food and the white body genetically capable of digesting it" (11). Comparing cow's milk with human breast milk, Andrea Wiley (2011) explains that cow's milk must be fortified in order to offer Vitamin D, and has "more protein, minerals (except iron), and some B vitamins, and less sugar, Vitamin C, and Vitamin A" (9). Although the dairy industry promotes milk as a major source of calcium, a necessary mineral for growth and strong bones, the majority of the world's human population cannot digest cows' milk, and the claim that this milk produces strong bones and taller children is simply unsupported by the research to date (Wiley, 64–82). As of 2008, U.S. cows' milk consumption has declined to just 76% of what it was in 1970, while cows' milk consumption has increased 17 times in China and 2.4 times in India; both are developing countries where there has been a general rise in the demand for animal products as a sign of modernity and affluence (Wiley 91).

Animal studies scholarship includes the varying approaches of posthumanism, human-animal studies, and critical animal studies, all offering a range of perspectives for addressing milk across species, though this potential remains largely untapped. A recent article in the Human-Animal Studies journal, *Society & Animals,* appears to invoke Donna Haraway's posthumanist construction of factory farmed animals as "workers" in its framing and discussion of dairy cows as collaborating with the dairy producer (Porcher 2012). Haraway refers to "laboratory working animals" and "working animals in the food and fiber industries" as if being the experimental animal or the animal whose body is confined within the structures of industrial animal production were sustainable "jobs" that animals might willingly choose, or resign from (Potts & Haraway 2010). From the more activist standpoint of critical animal studies, "Haraway's work has become paradigmatic of a largely depoliticized approach within Animal Studies," evincing a clear humanist interest in human-animal relations that maintain species dominance (Weisberg 2009, 58).

To date, the majority of research on milk comes from Food Studies scholars, vegan feminist and ecofeminist scholars (Adams 1994; Adams & Donovan 1995; Gaard 1994, 2010b; Gruen 1993; jones 2004; Kemmerer 2011a). Feminist environmental science texts such as Sandra Steingraber's *Having Faith* (2001) and Maia Boswell-Penc's *Tainted Milk* (2006) address the nutritional value of human breast milk for mother and child, the "body-burden" of environmental toxins transferred through that milk, and more specifically the environmental racism challenged by Katsi Cook's Awkwesasne Mothers' Milk project. Advocacy groups such as Environmental Working Group and the MOMS and POPS project regularly monitor milk as an environmental indicator of health, and have found perchlorate (a rocket fuel) in every sample of California supermarket cows' milk (EWG 2004) as well as flame retardants (PBDE), pesticides (lindane, endosulfan, DDT), and other POPS in human breast milk ("Monitoring Mother Earth" 2009). This movement in environmental science affirms Katsi Cook's insight that the mother's body is the first environment, an insight that links the concerns of feminism, environmental justice, and interspecies justice.

Postcolonial studies offers another critical perspective, building on Alfred Crosby's (1986) concept of "ecological imperialism" to describe both the ruthless appropriation of indigenous land—particularly violating indigenous women (Smith 2005), queers and animals (Gaard 1997a)—and the introduction of exotic livestock and European agricultural practices (Huggan & Tiffin 2010). Ecofeminists Val Plumwood (2002) and Vandana Shiva (1997) have pointed out the ways dualistic thinking and instrumental reasoning of the "Master Model" have constructed nature, the indigenous, and the animal as "other" to meet human (elite male) needs, and biocolonialism functions as a continued practice, patenting indigenous knowledges and genetics, all under cover of "progress" through Western science and agribusiness. Environmental racism and classism exemplify additional contemporary colonial practices, linking the continued appropriation of *global* resources *with* the transfer of wastes to communities of color, and impoverished communities around the world. Until the work of Graham Huggan and Helen Tiffin (2010), postcolonial studies had yet to "resituate the species boundary and environmental concerns" at the center of its inquiry, examining the "interfaces between nature and culture, animal and human" (6).

Taken alone, each of these fields offers specific perspectives on knowledge while overlooking others. From a critical ecofeminist perspective, findings from these diverse perspectives can be brought together within a feminist framework that is grounded in awareness of the ecological necessity for gender justice across species, races, classes, and environments.

MILK MONEY: GIFT ECONOMIES VERSUS CAPITAL

In vernacular English, to "milk" something is to take it for everything you can get—but that's an adult's slang. For newborn mammals, mother's milk is a priceless gift: it offers nutrition, hydration, and affection, ecologically packaged at the right temperature. Breast milk helps to protect infants against common childhood diseases, ranging from diarrhea and pneumonia to respiratory tract infections, gastrointestinal infections, celiac disease, inflammatory bowel disease, obesity, diabetes, childhood leukemia and lymphoma, and Sudden Infant Death Syndrome (AAP 2012). Adults who were breastfed as children have lower blood pressure and lower cholesterol, lower rates of type-2 diabetes and obesity, and exhibit higher intelligence and stronger filial bonds of friendship and empathy (WHO 2012; Feldman 2012). Benefits for breastfeeding mothers include a quicker return to prepregnancy weight, temporary protection against conception, reduced risk of breast and ovarian cancer, and lower rates of obesity (WHO 2012).

But these benefits are not equally utilized by all populations across race and class, nor have they been researched in breastfeeding relationships that are commodified to cross race and class boundaries. For example, did the women enslaved or the wet nurses hired to breastfeed infants of the upper class enjoy the same biopsychosocial benefits as they would have in nursing their own offspring? Did the upper-class infant enjoy the same nutritional, emotional, intellectual benefits as s/he would have if nursed by the baby's birth mother? And while milk-sharing within women's communities has been both a traditional and contemporary practice, this sharing is part of a gift economy among voluntary participants. Can we describe as "milk-sharing" the nursing that takes places across species—as in a mother's voluntary and affectionate suckling of an infant of another species, as was common for pigs, dogs, monkeys, and cubs in precolonial Polynesia, the forests of South America, and the hunter-gatherer societies of Southeast Asia, Australia, and Tasmania (Simoons & Baldwin 1982)? Or is taking the milk of another mother—whether a human mother or a cow mother, goat, sheep, or elephant—to be appropriately described as "gift," "wages," or "theft"? What is milk "worth"?

In an editorial on the economic value of breast milk, *Mothering* magazine founder Peggy O'Mara (2012) did the math, bringing together the $4 billion a year in U.S. formula sales, the $1 billion annual health care cost-savings from breastfeeding, and the costs hospitals pay for handling donated breast milk—$50 per liter in Norway, $96–160 a quart in the United States. Using the American Academy of Pediatrics (AAP) recommended minimum of breastfeeding for the first six months of life,[1] the annual number of U.S. births (4,130,665), the amount of breast milk produced (25 ounces per day, or 140 quarts per 6 months), its value per quart ($96), and the percentage (13.3%) or number (549,378) of U.S. mothers who exclusively breastfeed at 6 months, O'Mara concluded: this small percentage of nursing mothers generates $7 billion in gross domestic product. In just six months, these breastfeeding moms outstrip the economic value of two years of formula sales (O'Mara 2012). Summarizing data from the AAP journal *Pediatrics*, the *Huffington Post* (Tanner 2010) places the value of breast milk even higher. Focusing largely on health care costs and savings, AAP pediatricians estimate the lives of 900 babies would be saved along with $13 billion in health care if 90% of U.S. women would breastfeed for the first 6 months of life, or $3.6 billion (in 2001) if only 50% of mothers breastfed. But today, only 12% of mothers follow AAP guidelines. Who are they, and why do the other 88% of moms stop nursing?

According to the U.S. Centers for Disease Control and Prevention (2010), there are racial and ethnic differences in breastfeeding, with educated middle-class Asian/Pacific Islander mothers breastfeeding at the highest rates (52%) at six months, followed by Hispanic mothers (45%), Euro-American mothers (43%), then American Indian mothers (37%) and African-American mothers (26%). Predictably, economic pressures make these percentages even lower. Histories of racism and colonialism in the United States legitimating the rape of indigenous and African women, the theft and sale of their children in boarding schools or in slavery, and the requisite nutritional and affectional neglect of African infants when their mothers were used as wet nurses and "mammies" for white slave owners' children all provide some historical context for today's low breastfeeding rates (Collins 1990; Smith 2005). For all women, the U.S. cultural hostility to the material realities of motherhood can be seen in the stigmas around welfare for single mothers and their children, workplace policies restricting maternity leave and flextime, women's persistently lower wages, and a lack of national legislation correcting these phenomena. According to Ann Crittendon (2001), becoming a mother is the most expensive workplace decision a woman can make, and "the single biggest risk factor for poverty in old age" (6).

Yet unpaid female caregiving is the lifeblood of families, and the heart of the economy. Including child-rearing, cooking, managing household finances, resolving emotional conflicts and chauffeuring, Edelman Financial Services estimated a mother's worth at $508,700 a year, not including retirement and health benefits (Crittendon 8). But in Crittendon's *The Price of Motherhood* and the book it inspired, *The Motherhood Manifesto* (Blades and Rowe-Finkbeiner 2006), breastfeeding gets less than two pages, and is always discussed in terms of maternity leave. Admittedly, the United States has the lowest rates of maternity leave of all industrialized nations, offering only twelve weeks of unpaid leave under the Family and Medical Leave Act (FMLA) of 1993, or six weeks under the California paid family leave law, as compared to Germany and Sweden's forty-seven weeks of full-time-equivalent paid leave, Norway's forty-four weeks, and Greece's thirty-four weeks; an international study of twenty-one nations' parental laws found the U.S. twentieth out of twenty-one, only one of two nations providing no paid leave at all (Ray, Gornick & Schmitt 2009). In addition to having to pay for these job-protected twelve weeks of parental leave, new mothers face other costly and cultural barriers to continued breastfeeding: the cost of a breastpump ($269 for Medela's Pump-In-Style electric breastpump), an office refrigerator to store the pumped milk, and workplace policies that include a lactation room and guaranteed breaks to use that room as needed.

Under these conditions, Crittendon calls the AAP breastfeeding guidelines "a sick joke" (258). As she explains,

In economics, a 'free rider' is someone who benefits from a good without contributing to its provision: in other words, someone who gets something for nothing. By that definition, both the family and the global economy are classic examples of free riding. Both are dependent on female caregivers who offer their labor in return for little or no compensation. (Crittendon 9)

Women's breast milk and women's labor are part of the gift economy that is simultaneously invisible, unmonetized, and appropriated in national and international economic systems. In Africa and Latin America, village women will share in nursing to relieve other mothers to work, or to support an infant whose mother is ill, has no breast milk, or has died (Waring 1988). In the United States, "lactivist" mothers have formed milk-sharing networks such as Facebook's "Eats-On-Feets" page that allows mothers who need breast milk and mothers willing to donate excess breast milk to find each other. This network uses the four principles for safe breast milk sharing—informed choice, donor screening, safe handling, and home pasteurization (Walker & Armstrong 2012)—and proudly contrasts their gift economy with the costs of milk banks, which may charge $4.50 an ounce for handling and screening donated (i.e., free) breast milk.

When women's breast milk is introduced as a market commodity, it fares poorly. In 2010, New York Chef Daniel Angerer produced his wife's breast milk cheese at Klee Brasserie and was promptly shut down by the New York Health Department (Inbar 2010). A year later, London's *Daily Mail* (2011) reported that a Covent Garden store, Icecreamists, had begun selling human breast milk in a champagne glass and labeled the product "Baby Gaga." Allegedly the woman who donated the first thirty ounces of breast milk was not paid for her time or bodily fluids, but she did have to undergo health screening; thirteen more women had volunteered to donate their breast milk as well. Reporting on his experience of eating human breast milk cheese in the *Village Voice* blogs, Robert Sietsema (2011) reported "it feels like cannibalism," (a sentiment echoed in several other posts online) and enumerated concerns that seem representative of those expressed on the blogosphere: "human instinct" says "there's something fundamentally disgusting" about it; excess breast milk should be donated for the nourishment of premature and critically ill babies; no one knows the effects of human breast milk on adults; and human breast milk products have not undergone the medical testing regularly used to screen cows' milk. But from a critical ecofeminist standpoint, Sietsema's final concern was most salient:

Women are not farm animals. Human-breast-milk cheese casts them in that role. There is nothing "ethical" about milking humans. What woman would consent to being milked for the culinary pleasure of others, unless strapped for cash? The natural result of this happening on a large scale is the exploitation of poor mothers, who will be tempted to sell milk and feed their babies formula.

Clearly, Sietsema's remarks rely on the human/animal divide. If eating women's breast milk "feels like cannibalism," what does it feel like to eat other females' milk? And what does it feel like to be a farm animal?

In a word, it probably feels like death—otherwise called "herd retirement." In February 2012, the Twin Cities' *City Pages* ran an article exposing "Cooperatives Working Together—a collective of America's biggest dairy co-ops, including Arden-Hills based Land O'Lakes—herd retirement program that slaughtered more than 500,000 dairy cows between 2003 and 2010" to raise the price of milk (Mannix & Mullen 9). The program worked: the dairy industry profited over $11.7 billion off herd retirement, raising prices for American dairy consumers in 2011 and driving more small family farmers out of business. Whereas in the 1980s, there were at least 8,500 small dairy farms in Minnesota, by 2007, that number had dropped to 2,000. Journalists Manning and Mullen profiled some of those small farmers, like Joe Sonneker, whose grandfather cleared their 160-acre lot over a century ago and started the small dairy farm he hoped to pass down for generations. In 2012, with Cooperatives Working Together controlling more than 70% of U.S. milk production, small farmers like Sonneker feel they don't have a voice in the future direction of dairying.

Of course, neither do the cows. Since the Depression, dairy cows have been producing more milk for humans than the market could handle, and when government programs to purchase and store the excess couldn't keep up with the costs, the U.S. Department of Agriculture (USDA) instituted a Dairy Diversion Program encouraging farmers to slow down production. When that effort failed, the Dairy Termination Program was instituted, encouraging farmers to sell their herds and get out of dairying for at least five years—but then farmers not involved in the program simply increased their herd sizes and production output. East Dublin Dairy in Murdock, Minnesota, is a case in point, milking over 5,280 cows twice a day. As the Pew Commission Report on Industrial Farm Animal Production (2008) documents, the current "efficiencies" in farm production that have arisen over the past fifty years are not sustainable, and their operations create an "unacceptable level of risk to public health and damage to the environment, as well as unnecessary harm to [food] animals" (viii).

For an animal whose natural lifespan is twenty to twenty-five years, cows in dairy production now survive only four to five years. The cow's milk output has increased from 2000 pounds/year in 1950 to up to 50,000 pounds of milk in 2004, a consequence of bovine growth hormones, that put enormous pressure on the cows' bodies (Keon 192–196). Artificially inseminated at fifteen months of age, a dairy cow suffers an endless cycle of pregnancy and lactation, milked two to three times daily by electronic milking machines, conditions that cause mastitis and other infections that must be treated with antibiotics. Fed an energy-dense food, she may spend her whole life confined in a concrete stall or standing on a slatted metal floor. Her calves are taken from her within hours after birth, with females kept to replace their mothers in the dairy, and males sent to veal farms, where they are confined in crates so tight they cannot move, and fed an iron-deficient diet until they are slaughtered at fourteen to seventeen weeks of age (Robbins 112–117). Predictably, the larger dairies also produce more manure and methane, polluting the air and water. Using Crittendon's critique of unpaid caregiving, the industrialized dairy system is also a "free rider," profiting at the expense of the cows, the small dairy farmers, and the dairy consumers as well.

Replacing breastfeeding's gift economy and severing the nursing relationship between mammal mothers and offspring, the industrialized dairy system of extracting wealth from animal nature, from labor and consumers, and concentrating it in the hands of the producer-owners is not "cooperative": in dairying, that term is now a Trojan Horse, concealing the ideological character of economics. To unmask its operation, a critical ecofeminist perspective is needed.

MOTHER DAIRY, MOTHER NATURE

Most westerners will recall Nestle's powdered milk campaign in Africa and India that persuaded thousands of young mothers to use powdered milk and infant formulas instead of their own breast milk, and thereby made corporate profits at the expense of widespread infant suffering, causing diarrhea, malnutrition and death. As documented by the British NGO War on Want, Nestle's baby food sales representatives *dressed like nurses* to give an appearance of scientific credibility to their sales in the poorer countries of Africa, Latin America, and Asia, including India (Alvares 1985). Due to poverty, lack of education, and lack of adequate facilities, many mothers in these countries could not read the instructions on the formula package, and did not have access to baby bottle sterilizing equipment or clean water. Instead, they put faith in the ideology of progress, and the superiority of technologically advanced nations: in a colonial world, indigenous people are pressured to share the viewpoint of the colonizer, to believe themselves inferior, and to adopt the ways of the colonizer in order to "improve." In India, multinational corporations like Nestle and Glaxo were criticized by the WHO for selling infant formulas and powdered milk, and an International Code for the Marketing of Breast-milk Substitutes was issued in 1981. Under cover of this international rebuke, an Indian national dairy corporation was quietly picking up Nestle's lost share of infant milk food sales. The story of Amul corporation and its engineering of India's Operation Flood is a story of third world elites joining first world corporations in colonial practices, with devastating effects on mothers and children, cows and calves, rural poor and small dairy farmers—a story that both parallels and exponentially magnifies the harms done to dairy farmers in the United States.[2]

Launched in 1970 and implemented in three phases until 1996, when European dairy food aid supplies ended, Operation Flood was the invention of Verghese Kurien, initiated as a solution to a difficult market situation (Alvares 1985). In the late 1960s, the European Economic Community (EEC) had a huge dairy surplus in the form of milk powder and butter, and after reconstituting some quantities and dumping others, the EEC finally sought to dispose of these products to the third world in the form of food aid. As Frances Moore Lappé and Joseph Collins (1977) explain in *Food First*, food aid has always been a colonialist extension of foreign policy, farm interests, and corporate interests; it is offered to open future markets for commercial sales, extending the reach of agribusiness corporations and enabling first world governments and economic institutions to control their third world counterparts. Lappé and Collins' theory well describes the outcomes of India's Operation Flood, through the wealth and rise to power of Verghese Kurien and the Amul Dairy Cooperative.

At the time of the EEC surplus, Kurien was a twenty-year employee of Amul, India's largest manufacturer of milk powder and butter; the food aid would ruin Amul's markets. But as chair of India's National Dairy Development Board (NDDB), Kurien was well positioned to orchestrate a solution. In his chapter on "Imperialism Through Food Aid: The Role of Third World Elites," Claude Alvares (1985) explains: Operation Flood would not use the food aid as charity offered for direct consumption by the poor; rather, "food aid would be sold to the public, and the funds generated [would be] invested for the long term dairy development of the country" (3). Dairying would be an instrument of *progress*, business elites argued, as small and landless farmers would be organized into cooperatives for enhancing milk production and negotiate better rates for their products. India's government saw that the project would generate funds for dairying that the government could not raise, along with providing milk for the middle classes in the cities and improving the economic condition of the poor. Accordingly, the Indian Dairy Corporation (IDC) was established in 1970 to administer Operation Flood and Verghese Kurien was appointed its chair.

The operations and proclaimed outcomes of Operation Flood were strongly criticized by development scholars Bharat Dogra, Claude Alvares, and Shanti George. In his famous exposé of Operation Flood, "The White Lie," first published in 1983 and denied via media and statistical manipulations by India's National Dairy Development Board and Indian Dairy Corporation, Alvares listed the actual outcomes of Operation Flood:

1. it created four 'Mother Dairies' (in Bombay, Calcutta, Delhi, and Madras), milk-processing plants that recombine solid milk products and butter oil into milk for city people;
2. it built a national milk grid radiating from Anand to transport milk from the dairy-processing plants, throwing local producers around these cities out of employment;
3. it made Amul the largest baby food producer in India, and the strongest opponent of the WHO's Code against advertising baby foods;
4. it diverted large stocks of imported commodities from the cities to Gujarat dairies controlled by Amul;
5. it made Kurien a consultant for the World Bank, overseeing a new third world expansion of Operation Flood in Bangladesh, Pakistan, the Philippines, Sri Lanka and China (Alvares 1985).

While the dairy cooperatives were already collecting more milk than they could sell, and converting the surplus into baby food and butter, in the rural town of Kaira, Indian farmers were depriving their own undernourished children of cows' milk in order to sell to the cooperatives; we aren't told why their human mothers weren't breastfeeding, or why mothers' milk wasn't sufficient, though poverty and malnourishment seem likely explanations. Alvares simply reports that these Kairan farmers barely got a remunerative price for their cows' milk.

Part of the problem had to do with the cattle: Operation Flood involved the import of European bulls, heifers, and crossbreeds. In India, cattle have multiple uses in agricultural tasks (traction, fuel, fertilizer) and though the indigenous cows and buffalo are poor milk producers compared to the exotic Northern European breeds, their upkeep is minimal, and they are environmentally sustainable. While the imported breeds require special feeds and are subject to diseases that need veterinary attention, indigenous breeds subsist on local vegetation and are adapted to the climate, withstanding diseases and parasites, and calving easily without human assistance (Sainath 2012). But the value of their subsistence milk could not be converted into profits for Amul Dairy. Fifteen years after the third and final phase of Operation Flood ended in 1996, Kerala's indigenous cattle population had declined by 48% (Sainath 2012).

Rural women have also been harmed by Operation Flood, as the production and sale of ghee, along with its economic returns, used to be the sole province of women. With Operation Flood, the new crossbreeds required additional feeding and milking labor from women and children, and the milk was sold for cash, leaving women no economic returns and lowering their status in the family economy. Alvares cites an Indian Council Social Science study on the impact of Amul Cooperative on women:

The enormous structure of the Amul complex at Anand, with a highly modern campus of steel frame, mosaic and glass, air-conditioned buildings, laboratories, gleaming aluminum and steel plants, white uniformed and capped staff, beautifully laid out gardens, sound proofed and plus seated auditoria and air-conditioned luxury buses seem very far removed from the lives of the village women whose work has made this glossy new world possible, but to which they have no entry. Not one of them has acquired mastery over the new technology that has taken over their traditional tasks of making butter and cheese for the urban consumer. They are not even aware that they are contributors to a development miracle that is assuming the size of a national movement. (37–38)

Despite these social costs, Operation Flood was celebrated among social and international elites. Bruce Scholten's *India's White Revolution* (2010) is a single-authored volume that prominently features a jacket endorsement from Kurien, copious footnotes and quotations of Kurien, and most poignant of all, a photograph of the Amul range of dairy products, including the tiny tin of infant milk food substitute with the picture of a fat, smiling baby. But even Scholten concedes to critic Shanti George when she points out that Operation Flood's "modernisations resulted in a net loss of women's status" as only men were employed in the new high-tech infrastructure (233). Even after retirement from NDDB chair, Kurien remained in charge of the Gujarat Coop Milk Marketing Federation, where he had already exported Amul butter and cheese to over forty countries, including the United States. India's "White Revolution" companioned its Green Revolution (Shiva 1997) in the colonial pattern of shifting subsistence production into cash commodities for export, thereby destabilizing an already precarious subsistence economy (often powered by women's work) and throwing thousands of people into real material poverty. In his book's conclusion, Scholten reports that Kurien has been contacted by "African countries such as Kenya, Uganda, Ethiopia, Mozambique and Rwanda" all expressing interest in replicating Operation Flood (254). Such expansion may displace nomadic cattle herders such as the Maasai, whose subsistence lifestyle is well suited to their environment; for other areas, an African Operation Flood will surely affect human health and nutrition if populations have no historic relationship with cattle herding, and thus have inherited no lactase for digesting milk beyond childhood.

MILK HAS SOMETHING FOR SOME BODIES

In the 1970s, the Washington, DC-based Physicians Committee for Responsible Medicine (PCRM) filed several petitions with the Federal Trade Commission (FTC) against the American Dairy Council, showing data proving that advertisements promoting dairy products violated federal advertising guidelines; accordingly, in 1974, the FTC's complaint of "false, misleading, and deceptive" advertising forced the Dairy Council to change their slogan from "Every Body Needs Milk" to "Milk Has Something for Everybody" (Keon 2010, 13). Both vegan milk critics (Campbell & Campbell 2006; Keon 2010; Oski 1977; Robbins 1987) and some food studies scholars (DuPuis 2002; Wiley 2011) have challenged the dairy industry's "perfect food" myth about milk. Vegan studies are more likely to emphasize the linkages between animal-based diets and many Western diseases such as heart disease, obesity, and cancers of the colon, breast, and prostate, while both food studies and vegan studies scholars concur in observing the association between osteoporosis and animal-protein based diets (Keon 2010, 173; Wiley 2011, 80), an association of greatest concern to women, whose hip-fracture rates are regularly double that of men's. These correlations are reinforced by the fact that "as [most notably Asian] populations move to a more Western, industrialized lifestyle, which often includes dairy consumption, the risk of osteoporosis increases" (Wiley 80; cf. Campbell & Campbell 2006). These scholars also agree on the eurocentrism and racism of the U.S. dairy industry's claims for the universal healthfulness of milk.

Populations that have a historic practice of milking domestic animals (Central and Northern Europeans, countries colonized by Europeans [the United States, Canada, Australia, New Zealand], and Saharan nomads) have retained the enzyme (lactase) that digests lactose sugar in milk, far beyond childhood; however, the majority of the world's populations lose the lactase enzyme by the age of four, and thus lactose intolerance is common among Vietnamese, Thai, Japanese, Arabs, Israeli Jews, and African Americans, Native Americans, Asian Americans and Hispanic Americans (DuPuis 27; Keon 45; Wiley 24). Rather than acknowledging this diversity in digestive capacities, the U.S. dairy industry has coined the terms "lactase impersistence" and "lactose maldigesters," terms that effectively pathologize non-white populations. In 1999, the PCRM again challenged the pro-milk agenda of the U.S. National Dairy Council, this time in the *Journal of the National Medical Association*, a publication serving African-American health practitioners, and in 2005 filed a class-action lawsuit against grocery stores and dairies in the Maryland and Washington, DC, area calling for milk carton labeling. The campaign was publicized by images of people of color clutching their stomachs, or doubled over outside a unisex bathroom, and captioned "got lactose intolerance? 75% of people do, particularly people of color. If you're lactose intolerant, you may have grounds for a lawsuit" (Wiley 32). Such grounds of racism in milk promotions have historical precedent. Food science scholars Andrea Wiley and Melanie DuPuis each quote histories of milk written in 1929 and 1933, respectively, to illustrate the precedence and persistence of overt racism from Depression-era claims for milk's capacity to produce racial superiority:

> The races which have always subsisted on liberal milk diets are the ones who have made history and who have contributed the most to the advancement of civilization. As was well said by Herbert Hoover in an address on the milk industry delivered before the World's Dairy Congress in 1923, "Upon this industry, more than any other of the food industries, depends not alone the problem of public health, but there depends upon it the very growth and virility of the white races." (Wiley 33–34)
>
> A casual look at the races of people seems to show that those using much milk are the strongest physically and mentally, and the most enduring of the peoples of the world. Of all races, the Aryans seem to have been the heaviest drinkers of milk and the greatest users of butter and cheese, a fact that may in part account for the quick and high development of this division of human beings. (DuPuis 117–18)

Continuing the theme of white power, patriotic milk promotion ads in the United States during World War II labeled a factory photo of workers on a milk-bottling line as "white ammunition" (Wiley 59).

Why has the dairy industry not been held accountable for its blatant ethnocentrism? Beyond a nexus of cultural hegemony and economics, one source (Campbell & Campbell 2006) points out the conflict of interest caused by allowing the same person to chair the National Academy of Sciences' Food and Nutrition Board, consult with several dairy-related companies (i.e., the National Dairy Council, Nestlé Company, and Dannon), all the while chairing the Dietary Guidelines Committee that established the Food Guide Pyramid and setting national nutrition policy affecting the National School Lunch and Breakfast programs, the Food Stamp Program, and the Women, Infant, and Children Supplemental Feeding Program (312). Such antidemocratic alliances across government, science, and industry appear to persist in both first-world and third world contexts, exemplifying the links between intranational and international colonization.

MATERIAL PERSPECTIVES ON MILK
AND BOVINE AGENCY

What is the embodied experience of a dairy cow, and how can we know it? To date, this question has been addressed primarily from the standpoint of the Animal Sciences—that is, those who unabashedly explore lactation, maternal behaviors, weaning distress, and the implications of breaking mother-calf social bonds in their research, all for the purposes of human profit. Some of these studies combine an animal welfare approach with their quest for profits,[3] while others seem purely production focused. Acknowledging the well-known role of the hormone Oxytocin (OT) in pregnancy, birth, and lactation, these milk studies examine OT specifically in terms of milk production and maternal behaviors, but seem surprised to discover that "the parallels between the findings in animals and humans are indeed remarkable" (Editorial, *Hormones and Behavior* 2012). One study compares milk production when mother cows are milked in the presence of their calves, and then allowed to nurse their calves, versus cows that are exclusively machine milked without their calves (Kaskous et al 2005); this study finds that the bodies of mother cows release more Oxytocin in the presence of their calves, and thus they produce more milk even though they nurse the calves after machine milking. Without their calves, mother cows produce little or no Oxytocin, reducing the milk production ("ejection") to such an extent that dairy farmers regularly rely on "tactile teat stimulation, either manually or by the milking machine," and dairy scientists believe "it is necessary to elevate oxytocin blood concentrations either by exogenous oxytocin or by applying nervous stimuli such as vaginal stimulation which are strong enough to induce endogenous oxytocin release" (Bruckmaier 2005, 271).[4] Another study explored the effects of oxytocin injected into cows whose milk production is disturbed by being "switched from suckling to machine milking" or by being "milked in unfamiliar surroundings" (Belo & Bruckmaier 2010). The study acknowledges that "in dairy practice, OT treatments are frequently applied intramuscularly at a very high dosage" (63) which increases oxytocin for a few hours, but has the lasting damage of desensitizing the cow's udder and producing a reliance on repeated injections to obtain milk.

Studies of "weaning distress" find that this distress can be reduced by "disentangling" the various aspects of weaning—cow-calf separation and the act of nursing (Jasper et al. 2008; Weary et al. 2008). Acknowledging that oxytocin is involved in nursing for both the mother cow (as a response to the presence of the calf and teat stimulation, OT promotes lactation) and her calf—OT is released in the calf's body "only when calves were nursing from the cow and not when drinking milk from a bucket"(Weary et al., 29)—animal science researchers then propose that OT is comparable to other "opiate-like substances in milk" rather than a material produced through and reinforcing attachment—and thus "young mammals develop an addiction to milk and without that source of opiates, they become like addicts craving their drug of choice!" (Weary et al., 29; cf. Newberry & Swanson 2008). Pathologizing oxytocin—the biological foundation of the mammal mother-infant affectionate attachment, a material and relationship crucial to species survival—animal scientists strive to construct their own role in separating mother cow-newborn calf dyads as simply hastening an act of healthy separation.

A cornerstone of animal science scholars' arguments is the theory of "parent-offspring conflict" first described in Robert Trivers (1974) and persistently cited as a fact supporting the commercial dairy farmers' practice of separating mother cows and calves within two to six hours after birth. According to this theory, "weaning conflict" arises from the fact that while mammal infants benefit from continued mothering and nursing, the "level of maternal investment" decreases with age, and the mammal mothers "do better" or "benefit" from investing in future reproduction and new offspring, leaving the older offspring to forage for themselves (von Keyserlingk & Weary 2007; Weary et al. 2008). Exploring variations on the timing of mother-calf separation, watering-down milk and providing sucking substitutes, and even comparing Harry Harlow's well-known abusive research on infant monkeys (Harlow & Harlow 1962) with human toddlers' use of stuffed animals to support the claim that "animals—including humans—*routinely* develop attachments to inanimate objects" (Jasper et al., 142; italics mine), animal scientists use false analogies and flawed logic in their attempts to produce scientific legitimation for the exploitive practices of commercial dairying. But they stand on slippery ground: all their data point to the fact that severing the mother cow-calf relationship—a complex relationship that involves a constellation of maternal behaviors responding to the copresence of mother and calf (licking, sniffing, nursing, calling, and biobehavioral synchrony)—is what causes emotional, behavioral, and biological distress.

Instead of focusing on the material of milk production, studies of oxytocin in human mammals tend to focus on relational behaviors, attachment, nurturance, empathy, and happiness—yet material and relational elements are present for *both* bovine and human mother-infant pairs. In the special 2012 issue of *Hormones and Behavior* addressing "Oxytocin, vasopressin and social behavior," scholars suggest that these hormones "modulate human social behavior and cognition," "enhance interpersonal trust," and "have been linked to attachment, generosity, and even pair bonding in humans" ("Editorial comment"). As with other animal studies, these studies of human relations showed that oxytocin release is behaviorally influenced, promoted by the mother's breastfeeding, her physical proximity (which includes touch, odor, movements, body rhythms), her affectionate gaze and vocalizations, which together create the biobehavioral synchrony that lays the foundation for such social, emotional, and cognitive competencies as self-regulation, empathy, social adaptation, and a reduced risk of depression, as well as supporting more secure romantic relationships in adulthood (Feldman 2012). Mothers' oxytocin response is moderated by contextual factors and individual characteristics (Strathearn et al. 2012); it may be influenced by the mothers' experiences of parental care from their own childhoods, thus paralleling research for other animal species that indicates the generational transmission of OT operates through parenting behaviors (Feldman 2012).[5] Most interesting for gender equity is the study assessing oxytocin levels in first-time parents, which found "comparable levels of baseline OT in fathers and mothers," indicating that "active paternal care provides one pathway to activate the OT system in bi-parental mammals, which in mothers is triggered by birth and lactation" (Feldman, 385). In this study, both parents showed biobehavioral synchrony with their infants, though each parent engaged in gender-specific behaviors (i.e., fathers tended to throw the infant in the air, move the child across the room, or move the child's limbs, behaviors which nevertheless increased the fathers' OT levels).[6]

Animal science research can thus be used to undermine or to advance animal industry and technology, and influence interspecies relations. The dance of infant cry and maternal milk letdown is biologically and behaviorally encoded, but the code can be broken; in 2012, animal scientists are selectively breeding cows who seem indifferent to separation from their newborn calves. "How do you break a wild animal?" asks pattrice jones. "The key can be found in the word itself: You sever connections" (jones 2006: 321).

Inside each glass of milk is the story of a nursing mother separated from her offspring.[7] To justify and feel comfortable in "breaking" the biopsychosocial bonds that join mother and calf, dairy scientists, dairy farmers, and dairy consumers alike must deny the web of relationships that defines healthy ecosystems. Although animal science scholarship provides ample documentation of the distress this separation produces for both mother cow and calf—"vocalizations" averaging more than 120 calls during 20 minutes for the calf (Jasper et al. 2007), and "increases in vocalizations and activity" for the mother cows (von Keyserlingk & Weary 2007)—the abstractions of the words used to describe this distress shield us from the images of the cows and calves themselves. Bovine resistance to commercial milk production is concealed in these animal science studies, and requires a critical animal studies approach to uncover.

Animal activists confirm that cows separated from their calves bellow and appear to grieve for days afterward, sometimes ramming themselves against their stalls in attempts to reunite with their calves. News articles report the "amazing" feats of cows returning across miles of countryside in order to nurse calves from whom they were forcibly separated (Dawn 2008, 162–64). Some cows even use subterfuge to deceive dairy farmers and protect their calves. Veterinarian Holly Cheever (2012) recounts one such experience when she was contacted for consultation by a dairy farmer whose cow was mysteriously dry. With her fifth pregnancy, the cow had disappeared to give birth and returned with her calf, which the dairy farmer promptly removed; she was milked morning and night, but produced no milk. Days later, the farmer called back: he had followed the cow out to pasture during the day and discovered her secret. The mother cow had given birth to twins, and had hidden one in the tall grasses. As animal science researchers acknowledge, "under natural conditions cattle 'hide' their young away from the herd, returning at infrequent intervals during the day to suckle" (von Keyserlingk & Weary 2007, 108). But what Cheever noted is a sophisticated conceptual process that the dairy scientists didn't predict: this cow was capable of remembering the four prior births and the loss of those calves; this cow was capable of anticipating a similar fate for her new offspring; this cow made a kind of "Sophie's Choice" decision in choosing which of the twins to bring back to the dairy farmer, and which of the twins to hide and protect; this cow was capable of subterfuge, stealthily returning to nurse her newborn each day, then presenting herself for milking at the usual hours, morning and night. Though Cheever (2012) "pleaded for the farmer to keep her and her bull calf together, she lost this baby, too—off to the hell of the veal crate." Cheever's observation documents an example of farmed animal agency and resistance.

How does drinking this bovine mother's milk shape human identity? Who do we become?

CRITICAL ECOFEMINIST MILK STUDIES

In California and Wisconsin, rows of cows are lined up in stalls, with metal suction cups pumping on their teats, extracting milk; on the May 21, 2012, cover of *Time* magazine, a twenty-six-year-old mother is pictured, breastfeeding her three-year-old son (Pickert 2012). Which image is more shocking?

Ideologically imprisoned in a humanist colonial framework, few human mothers who breastfeed their infants use this embodied experience as an avenue for empathizing with other mammal mothers; few human parents who touch and nurture their newborns have used these behaviors' affectionate oxytocin release as an opportunity to consider the experiences of other animal parents locked in systems of human captivity. A critical ecofeminist milk studies addresses the biopsychosocial connections produced through the behavioral and material elements of this first relationship, the mother-infant bond, and their *nursing milk* (Kemmerer 2011a).

For too long, the dominant culture has childishly projected its own gendered image onto nature as selfless and self-sacrificing mother, as in Shel Silverstein's *The Giving Tree*, or onto other mammal species, requiring the female bovine to symbolize maternal nature: mindless, patient, slow-moving, lactating. If we set aside this stereotype and look into her eyes, what can we see?

NOTES

1. The discrepancy between the World Health Organization's (WHO) recommended two years of breastfeeding and the AAP's recommended six months has more to do with cultural and economic contexts than it does with babies. As the WHO confirms, breast milk contributes to infant health and emotional well-being on multiple levels, providing 100% of infant nutrition for the first six months of life, up to half or more of nutritional needs for the next six months, and up to a third of nutritional needs the second year of life (see http://www.who.int/nutrition/topics/exclusive_breastfeeding/en/). In global contexts, and particularly in rural or developing countries, infants who are fed formula are at greater risk of mortality due to unsanitary conditions (i.e., polluted water, unwashed bottles, diluted formula, etc.) and thus breastfeeding is a greater protection for life; moreover, a culture of prolonged breastfeeding persists in less industrialized parts of the world. In a first-world context, where mothers are more likely to have access to bottle sanitization, purified water, economic or food aid and infant formula—coupled with the pressure to earn income shortly after childbirth, and the heteropatriarchal sexualization of women's breasts as toys for adult men rather than as functional sustenance for infants—the AAP strategically recommends the minimum duration for breastfeeding, and yet only 12% of mothers meet even this recommended minimum, a sharp decline from the 70% of mothers who breastfed at the beginning of the 1900s (see Wright and Schanler 2001). As I explain, the other 88% face barriers that are linked to larger systems of racism and classism; see "Racial and Ethnic Differences" at http://www.cdc.gov/mmwr/preview/mmwrhtml/mm5911a2.htm

2. Operation Flood has been called the "White Revolution" alluding to the Green Revolution of biotechnology and genetic engineering heralded by agricultural corporations such as Monsanto and Cargill, and strongly critiqued by scholars such as Vandana Shiva (1993, 1997) as a pseudorevolution involving the massive theft of indigenous knowledge, biodiversity, seeds, and genes.

3. Of the studies surveyed here, scholars associated with Animal Welfare Programs were universally housed in Animal Science and Food Science programs, not in the Humanities.

4. In this quotation, humanities scholars will readily note the passive voice, a mode of diction that neatly sidesteps the question, "who is stimulating this cow's teats and vagina, and for whose pleasure?" As pattrice jones (2011) has remarked, linking heterosexism and speciesism: "A primary tenet of gay liberation is that what consenting people do with each other's bodies is nobody else's business. And, of course, eating meat is something you do to somebody else's body without their consent" (47). Her observation could be applied equally well to this animal-science-initiated and uninvited sexual abuse of "dairy" cows.

5. Primate studies show that females deprived of socialization and maternal care during early life do not learn these maternal skills and are abusive toward infants, suggesting a bio-behavioral connection between oxytocin release and nurturing behaviors (Curley & Keverne 2005; Harlow & Harlow 1962).

6. Along with diverse brain regions, oxytocin is also "released at peripheral cites, including the heart, thymus, gastrointestinal tract, uterus, placenta, amnion, corpus luteum, and testes, underscoring the widely-distributed and dynamic nature of OT

production in body and brain"—confirming the biological basis for human males to release oxytocin in strong affiliative relationships (Feldman 2012, 382).

7. Lisa Kemmerer (2011a) proposes the term "nursing milk" which foregrounds the relational constitution of milk. To conceal this relational ontology, the U.S. Dairy Industry promotes advertising that presents milk as a commodity, an object of liquid-in-a-glass that can produce "milk mustaches" and "strong bones/bodies" while concealing the fate (veal for males, future dairy cows for females) of those calves for whom the mothers' milk was created to feed. No wonder that the viewing public has conveniently forgotten the fact that milk comes from teats and not cartons; such elision enables industrial dairy sales and production.

Chapter 4

Fireworks

And the rocket's red glare, the bombs bursting in air
Gave proof through the night that our flag was still there.

—"Star-Spangled Banner" lyrics by Francis Scott Key[1]

Each year on July 4, parks and waterfronts in the United States set off firework displays to commemorate Independence Day. Anticipating the evening's spectacle, families gather for picnics and leisure recreation, and some individuals set off small fireworks—sparklers, "snakes," spinners, rockets, and fountains. When the public fireworks finally begin to whistle and explode in the night sky, few spectators notice the startled geese flying erratically against the lights, the dogs breaking leashes or jumping fences to escape the explosive bangs and whistles, the infants and toddlers wailing and begging their parents to take them away while the parents insist they stay for the "fun." How do firework displays affect other animal species, and all those categorized as "other"—indigenous North Americans, for example, or recent refugees? Do the fireworks' explosions prompt some to re-experience the warfare suffered by their parents and ancestors? And where do these fireworks land—in lakes and rivers, in soil and air, trees and bushes? What do spectators know about the fireworks' manufacture, their purchase price, or funding sources? In the United States and in many countries, fireworks displays are promoted as a communal (and monolithic) act of festivity and all views to the contrary are "spoiling the fun."

Is everyone having fun at the fireworks? Interrogating the narrative of fireworks and constructing a critical ecofeminist perspective on this cultural phenomenon requires an interdisciplinary approach that utilizes history, refers to national and cultural contexts, and explores the role of art and science, militarism, religion, and politics. What and whose stories are being told in these firework celebrations? And whose stories, whose views, are excluded? How do fireworks displays construct the identity of the viewing audience? This investigation is made more difficult by the fact that "outside the literature of pyrotechnic manuals, the memory of fireworks has existed almost exclusively in the realm of ephemeral literature and imagery—in newspaper advertisements, periodical articles, handbills, pamphlets, trade cards, engravings, and popular prints" (Werrett, 246). Working with primary and secondary texts as well as the standard tools of feminist and ecocritical analysis, this chapter explores fireworks as a narrative, bridging science, art, and colonialism.

FIREWORKS IN EUROPEAN HISTORY

Allegedly invented by a Chinese chemist who mixed the fire-starters sulfur and charcoal with the propellant, potassium nitrate, early fireworks were first packed into closed bamboo tubes as early as the seventh century ("Modern Marvels"). Sometime thereafter, fireworks were brought from China to Italy and on to the rest of Europe. Clear textual documentation of fireworks displays exists for the fourteenth and fifteenth centuries, and is joined by visual documentation in the sixteenth and subsequent centuries; initially portrayed through etchings and engravings, fireworks later appeared in pen and ink drawings, watercolor paintings, gouache, oils, and lithograph (Salatino 1997). Reconstructing the development and displays of fireworks from these written and visual texts, Simon Werrett (2010) emphasizes the interactions between artisans and scientists in the productions of fireworks displays, while Kevin Salatino explores the distinction between the pyrotechnical event, and the images that recorded and narrated the event as spectacle.

Although European fireworks began with military pyrotechnics and returned to these in the nineteenth century, the heyday of "artificial fireworks" from the fourteenth to the eighteenth centuries required a collaboration of "gunners" (military fireworkers), courtiers, natural philosophers (i.e., alchemists, astronomers, physicists), architects, painters, and academicians in creating elaborate fireworks spectacles celebrating the power, authority, and divine appointment of the ruling elites. Detailing the development of fireworks spectacles in France, England, Russia, and Italy, Werrett (2010) shows how differing cultural and political contexts shaped firework displays and influenced their meanings. Londoners, for example, variously praised or condemned fireworks in conjunction with the rise and fall of anti-Catholic sentiments, and natural philosophers responded: when the public feared Jesuit incendiarism, natural philosophers eschewed making pyrotechnic associations with their experiments, but by the end of Queen Anne's reign in 1714, fireworks were seen as evidence of divine powers, and part of natural theology. In France, fireworks were intended to serve the king, and the "literature on fireworks presented the elements themselves, not artisans or academicians, working to please the king" (Werrett 237). In Russia, the new Academy of Sciences in St. Petersburg found it was only able to interest the nobility in the sciences via fireworks: while scientific lectures flourished in other national and cultural contexts, lecture programs were a failure in Russia, but fireworks allegories succeeded in arousing scientific interests among the nobility. And whereas the English fireworkers acknowledged the efforts of artisans and natural philosophers in their productions, in Russia, academicians strove to make artisanal labor invisible so that royal funding would be directed to the Academy of Sciences.

As Salatino's research confirms, "the propaganda value of these costly, ephemeral entertainments rested less on the event than on its offspring, the record of the event. The illustrated fete books and prints produced and disseminated throughout Europe were a far more effective means of promoting influence and authority" than the event alone (1997, 3). As early as 1637, for example, Ferdinand III requested that "each night programs be distributed" to accompany the fireworks celebrating his election as Holy Roman Emperor, anticipating that without clear instructions, the viewing populace was not sufficiently literate to understand the subtle allegories unfolding in the transformations of the *macchina* (fireworks structure, or "machine") (Salatino 19). By the middle of the eighteenth century, any fireworks display was accompanied by clarifying printed programs, otherwise, as one exasperated scholar put it, "the wit that one employs is a waste of time for the majority of the spectators" (Salatino, 19). But even these attempts to enforce fireworks propaganda were not able to stifle dissent: some writers expressed disappointment with the storms and excessive smoke from the 1770 fireworks at Versailles celebrating the marriage of Marie Antoinette and Louis XVI, and still others complained about the exorbitant costs, which (then and now) include mishaps, spectator deaths, and the redirection of funds away from the general public to the fireworks celebrations. As curator of a J. Paul Getty Museum special exhibition on fireworks, Salatino is careful in his accompanying monograph to emphasize that the images recording the firework spectacles are best understood in terms of their larger political, cultural, and historical context. Yet despite these varying contexts, certain features regarding the relationship between European cultures' pyrotechnics and manipulation of nature persist.

In the anonymous engraving from England, offering "A representation of the fire-works upon the river of Thames, over against Whitehall, at their majesties coronation" (Figure 4.1) produced for King James II in 1685, three columns of fireworks erupt from islands in the river: a Roman warrior stands on the left, complete with helmet and armor, and the inscription PATER PATRIA ("father of the country") above him; to the right, a crowned figure holds a serpent in one hand and four arrows in the other, while sitting astride a fire-breathing dragon, with two crowned figures pinned beneath its tail, and a mermaid under the dragon's head. MONARCHIA is inscribed above this figure. In the central column, the initials "JMR 2" are wreathed in laurel leaves beneath a crown with a cross, and above these, the sun. An unnaturally placid (in the presence of fireworks) line of swans, six adults and numerous offspring, swim in front of these flaming columns, pursuing a floating warrior who vanquishes the skeletal face of death. The union of military and monarchy in this fireworks coronation spectacle is presented as endorsed and guided by both nature (swans, sun, fire, water) and the divine (perhaps a tribute to King James' belief in the divine right of kings). Fireworks spectacles were an expensive and lengthy endeavor in European court life, persistently affirming the triumph of what E.M.W. Tillyard (1959) first described in *The Elizabethan World View*: a society ordered through hierarchy and specifically through monarchy, with the heteromale king at its center like the sun around which everything—society and nature—revolved.

Figure 4.1   A representation of the fire-works upon the river of Thames, over against Whitehall, at their majesties coronation (1685).

Luca Ciamberlano's triple engravings of fireworks for Emperor Ferdinand III's visit to Rome in February 1637 aptly illustrate the resonances of classism, monotheism, and imperialism. Over a period of three nights, the fireworks machine simulating Mount Etna (the tallest active volcano in Europe, located on the eastern coast of Sicily) went through a series of transformations, depicting a different triumph each night: first, the Holy Roman Empire's victory over its rebellious subjects, the German Protestant Princes; then, over heresy (Protestantism), and finally, over the Ottoman Turks (see Figure 4.2). Not only is the classism, monotheism, and imperialism relevant from a critical ecofeminist standpoint, but the fact that these subordinated subjects are represented by fire-breathing dragons, peacocks, gorgons, as well as rams, lions, antelope—in other words, *animals*—attests to the triumph of human culture over animal nature. Likewise, the elements themselves are at war—earth, air, fire, water—with fire becoming the triumphant symbol of the Emperor himself, personally credited with creating imperial order out of nature's chaos. Thus, the narrative of fireworks is used to *animalize* humans of a subordinate class, religion, or ethnicity, and to portray nature itself as internally divided and at war, with order possible only through dominance

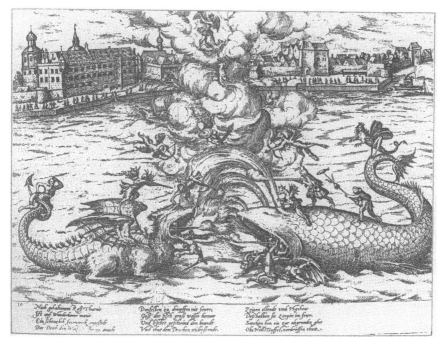

Figure 4.2   Franz Hogenberg, "Fireworks machines in the form of sea monsters on the Rhine at Dusseldorf" (1587).

and hierarchy. In short, militarist imperial propaganda was naturalized through the spectacles of fireworks.

This narrative spectacle linking military and monarchy is complemented and reinforced by species dominance. Franz Hogenberg's (1587) "Fireworks machines in the form of sea monsters on the Rhine at Dusseldorf" celebrates the marriage of the Duke of Julich by depicting warring sea creatures, a fire-breathing dragon, soldiers waving firearms, all struggling against an explosion of fireworks. Here, the trope of militarism and monarchy coexists uneasily with the alleged cause for celebration, love vanquishing war (see Figure 4.2). Similarly, in Charles-Nicolas Cohin le fils' "Fireworks and illuminations for the birthday of the dauphin" (1735), Hercules uses his fire club to slay the dragon against a backdrop of colonnades and fireworks, affirming human militaristic culture's triumph over animal nature and the reassertion of order via the monarchy (see Figure 4.3). As Simon Werrett observes, "in a world without electric light, fire was a powerful medium, a source of light and heat whose divine and magical connotations were strong" (3). Many social institutions and social climbers—from the Catholic Church to artillerymen; from courtiers

Figure 4.3   Charles-Nicholas Cochin et fils, "Fireworks and illuminations for the birthday of the dauphin" (1735).

to painters and architects; and from artists to natural philosophers and entrepreneurs—sought to take credit for the spectacle of pyrotechnics as evidence of their own mastery over nature.

By the middle of the eighteenth century, Edmund Burke's theory of the sublime influenced the production and interpretation of fireworks spectacles. Long associated with mountains, the sublime achieved even greater heights in Burke's 1757 treatise, *A Philosophical Enquiry into the Origin of Our Ideas of the Sublime and Beautiful*, where he established the mutual exclusivity of these terms: both encourage awe or veneration, Burke argued, but only the sublime induces horror. As Salatino explains, Burke's notion of the sublime as exciting "the ideas of pain, and danger" as well as terror, magnificence, profusion, and the sense of narrowly escaping injury or death all provide ideological context and shape to fireworks spectacles. Along with numerous examples of the fireworks *macchina* represented as explosive mountains, Salatino uncovers narratives of warfare-as-fireworks and fireworks-as-warfare that clarify the connections between militarism, fireworks, and the sublime domination of nature. Multiple sources acknowledge the associations between fireworks and warfare, both material (in the gunpowder) and metaphoric (fireworks bursting to delight spectators, or bombs bursting to maim, injure, and kill). As Salatino argues, the eighteenth-century concept of the sublime offers a perspective that sees war as an experience that can elicit pleasure when the experience resembles—but is not—warfare, and the explosions, bright lights, and bursting rockets offer spectators the experience of delight mixed with fear.

Fireworks spectacles of the eighteenth century used artifice to narrate the triumph of techno-science over nature particularly through comparisons to nature's fireworks (i.e., the eruptions of Vesuvius, famous for its destruction of Pompeii in 79 CE). At the papal Castel Sant'Angelo in Rome, the Catholic Church offered spectacular fireworks displays (called the Girandola) to celebrate the election of a new Pope, as well as drawing attention to significant holy days of the faith, Easter and the Feast of Saints Peter and Paul. While there seem to be many reproductions of these displays, Joseph Wright's paired images of the Girandola and of Vesuvius erupting are noteworthy because Wright "placed the volcano and the Castel Sant'Angelo in the corresponding part of each canvas (the right background), the one more effectively to mirror the other" (Salatino 49). In a letter, Wright compared these paintings, saying "the one is the greatest effect of Nature, the other of Art that I suppose can be" while a twenty-first-century art critic describes Wright's painting as "an almost apocalyptic vision" (King 2010) perhaps referencing Christianity's faith in a sublime apocalypse.

From a critical ecofeminist perspective, the eighteenth-century sublime does not further the goals of feminism, ecology, or democracy. Its valuation of transcendence over immanence, its pairing of pleasure and terror, and its celebration of spectacle over engagement, warfare as amusement, and the technological triumph over nature are all incompatible with an approach that values mutually-rewarding interconnections, relational ways of knowing, and ecological sustainability. As Patrick D. Murphy argues, this eighteenth-century conception of the sublime has affected the appreciation of nature external to the human body and external to civilization and ultimately "works against ecological values because it places a premium on the human mind separated from the body and the brain as a source of immaterial ideas" (2012, 90). Rethinking the sublime "requires a rethinking of the masculinist attitudes toward power and violence that seek out and infuse near-death events and reckless behavior with delight and an egotistical illusion of mastery" (92). Murphy's reading of the sublime underscores its ecological, ideological, and social liabilities.

In eighteenth-century Europe, Werrett's "geography of fireworks" changed as practitioners circulated throughout Europe, with Italian architects and artificers bringing new techniques and forms of display to the European courts. What began as a once-only nighttime spectacle grew to festivals of several days, performed on an elaborately built *macchina*. The cost of these spectacles eventually led to their demise in the latter half of the eighteenth century, and fireworks shifted from courtly to commercial contexts; the nineteenth century saw the rise of more commercial fireworks, appealing to more middle-class audiences. Class distinctions in firework displays are significant, as were the distinctions between the natural and the artificial. Class was maintained and performed largely via the ways spectators *managed and interpreted their own emotions*, as Werrett explains:

> Around 1600, when fireworks were relatively rare events, the court and the nobility claimed distinction from the vulgar through experience and knowledge of fireworks, which were otherwise terrifying spectacles, indistinguishable from natural portents. By the early eighteenth century, fireworks remained fearful, but now a wider audience of nobles approached them nonchalantly, manifesting control of the passions and the body through an artful apathy to dramatic spectacle. . . . *With the aesthetics of the sublime*, distinction from the vulgar was made to turn on enjoying fear rather than suppressing it, the expression of a pleasurable fear being considered *more natural* than the artifice of nonchalance. (239, italics mine)

Naturalizing the fear-pleasure association, the sublime aesthetic of fireworks socially constructs and narrates the experience of one's own bodily sensations, preparing fireworks spectators to accept what would otherwise be fear-inducing warfare as enjoyable, natural, even elite. Today, the persistence of this view can be seen in the repetitive and popular Hollywood films of the warfare genre—whether science fiction, espionage, enemy terrorist, historical, or futuristic—almost all of which employ firework-like special effects.

What happened to the large fireworks spectacles of the eighteenth century? By the nineteenth century, political economists had publicly portrayed fireworks as wasteful and unnecessary, middle-class pressure groups complained of the fireworks' dangers and vulgarity, and the era of vast pyrotechnic spectacle was replaced by smaller-scaled community fireworks celebrations, magic lantern shows, and the peep show or kaleidoscope. Unlike the actual fireworks display, Salatino observes, these optical reconstructions were "repeatable to the point of ennui" thereby reducing the sublime "to the merely curious" (97). But the aesthetics of the sublime persisted.

MODERN-DAY ANGLOPHONE
FIREWORKS IN CONTEXT

In the United States, fireworks are associated primarily with Independence Day, although they are also used at public New Year's celebrations, and at sporting events. Strong associations of fireworks and nationalist identities affirm the persistent, yet culturally contextualized variations of fireworks from Renaissance and eighteenth-century Europe.

In Northern Ireland, fireworks displays have multiple interpretations stemming from conflicts between Roman Catholics and Protestants, and their related desires for Irish nationalism or union with Great Britain. Fireworks are the most distinctive feature of Halloween celebrations, which range from "an essentially adolescent boys' culture surrounding bangers and squibs [small-scale fireworks] to family backgarden displays to large-scale public events" (Santino 1996, 215). Although fireworks have been banned since 1971 for all but official municipal uses, they are still widely available, and family fireworks often coexist with Halloween bonfires. Though many Irish insist that Halloween is a nonsectarian holiday where different traditions coexist in a celebratory way, it also "falls into the cultural property claimed by Roman Catholics," whereas Guy Fawkes Night (November 5) is "clearly both British and often anti-Papist, anti-Catholic" (Santino 1996, 228).[2] Both Guy Fawkes Night and Halloween have bonfires associated with them and are known as "Bonfire Night," their proximity in dates lending to some elision between the festivals. In Santino's and other studies, fireworks seem to be mapped onto and deployed as signifiers for existing social relations, cultures, and histories; they do not seem to be adding to a cultural lexicon, but rather illuminating one that already exists.

In Canada, fireworks have been one of the most popular and most expensive elements of festivities marking Dominion Day (when the Dominion of

Canada with its four original provinces was created on July 1, 1867) (Hayday 2010). For the first fifty years, the Dominion Day holiday was celebrated with picnics and fireworks organized by community groups and municipalities. But after the Second World War, the federal government became interested in using Dominion Day as an opportunity for nation-building and identity construction. In 1958, the first nationally televised broadcast of Dominion Day festivities at Parliament Hill included a twenty-one gun salute, trooping of the color guard, and fireworks. An official communication suggests that "the organization of an annual public festival on July 1 would establish in the memories of present day children the happy memories which their parents and grandparents have of May 24 [the former Victoria Day]" writes Matthew Hayday, explaining that the festivals were a conscious effort to address "two key target groups: children and immigrants" (2010, 295). Although Hayday does not focus exclusively on the fireworks displays, his account offers important images of national identity construction via these celebrations. Whereas the first Dominion Day in 1958 described "two great streams" influencing Canadian history, the English and the French, and featured an assimilationist performance from the Cariboo Indian Girls Pipe Band playing bagpipe music, the holiday was renamed Canada Day in 1982, and by 1992, event organizers crafted messages focused on "citizenship, official languages, the environment, and symbols of Canada" that would appeal to "youth, Natives, and multicultural organizations" (Hayday 2010, 309). By 2010, "hundreds of thousands of people flood Ottawa's streets every 1 July" and "in communities across Canada families attend local picnics and flock to firework displays" (313). As Hayday's research confirms, fireworks in Canada (as in the United States) are used as a tool within a larger cultural narrative of nation-building and citizenship identity construction.

In her study of nineteenth-century children's books and their presentation of New England's Independence Day festivals, Lorinda Cohoon notes the social construction of childhood citizenship as particularly inflected by race, class, gender, and ethnicity. In an era when neither white women nor African-Americans of any gender were allowed to vote, children's periodicals depicted strong connections "between nation and constructions of [white] masculinity . . . [particularly] boys' participation in Fourth of July events" (2006, 133). Although the direct witnessing and participation in repeated cultural rituals associated with nationalism and independence does shape children's civic identities, like Salatino's claim for the propaganda value of written fireworks programs, Cohoon claims that the children's periodical served as another powerful cultural form because of its persistence. Children's periodicals of the era depict parades and Independence events that "uphold an Anglo-American version of independence, which focuses on the New England forefathers and the privileges their struggle granted to white, middle-class New England readers" (2006, 134). In parades and periodicals alike, the Fourth paradoxically "celebrates freedom while reinforcing the idea that freedom's privileges are most readily available to those who conform" (135): "the 'sons' of the new nation must celebrate their forefathers' rebellions, but they must also comply with the laws and regulations of their towns and nation" (136). The Independence episodes written for children also reflect the ongoing national struggles over slavery, westward expansion, and treatment of Native Americans; they make invisible the growing national debate over women's suffrage. At a time when the Indian Removal Acts of 1830 made it legal for settlers to take the lands and lives of indigenous people, children's periodicals used the figure of the child as a representative of the nation, contrasted with "racially marked children and those who are described as especially powerless and vulnerable metaphorically represent[ing] the nation's views on slaves, Native Americans, and also recent immigrants" (142). Cohoon argues persuasively that the narrative presentations of these Independence celebrations "provide young children with national narratives that make the embarkation upon expansionist endeavors possible" and "model exclusionary practices" that by the 1860s and 1870s become elements limiting suffrage and immigration (2006, 148).

From New England Protestants to Southern Baptists, fireworks festivals persist in constructing and celebrating a nationalist identity in the United States. In his essay on "Fireworking Down South," Brooks Blevins (2004) brings a cultural studies perspective to his own decade of experiences selling fireworks in the U.S. South, providing photos of fireworks shacks, film clips, album covers, and lyrics, Confederate paintings, European fireworks literature and art, and again, page after page of fireworks stands. Southerners have an affinity for fireworks surpassing any other cultural group, Blevins explains: in the United States, "we all *watch* fireworks. In the South, though, we *use* fireworks; it's a hands-on affair" (28). Today, seven of the former eleven Confederate states have some of the nation's most liberal fireworks laws, making the South the most "fireworks-tolerant" region in the United States; as Blevin wryly remarks, "among the litany of rights cherished in the South is the right to endanger oneself and anyone else who happens to be in the vicinity" (31). For over a century, Southerners used to shoot off fireworks on Christmas more often than on Independence Day, a practice stemming from "unrepentant Confederates" who refused to celebrate a Yankee holiday after the Civil War in the company of their former slaves; one southern white woman, Mary Chesnut, called that first Independence Day the "Black 4th of July" (28). Even in 2004, Blevins asserts, the Deep South can be delineated by finding where fireworks can be sold for Christmas. Along with fireworks' racially inflected meanings, there's a "noticeable socioeconomic dimension to the fireworks crowd," writes Blevins, comparing the fireworks industry to the lottery, with the most frequent consumers drawn from the lower socio-economic levels. Fireworks are popular in the South due to a combination of factors: "the Jeffersonian tradition of hands-off government . . . as well as the region's relative rurality and lack of a strong environmental movement—characteristics also true of the Plains states, which combine with the South to form a sort of 'fireworks belt' strapped across the midsection of the country" (29–30).

Whether in Renaissance Europe or nineteenth- and twentieth-century North America, fireworks have been used to affirm, celebrate, and illuminate culture's dominance over nature, aligning social elites with culture over and against social subordinates and their association with nature. Fireworks offer a narrative that has been manipulated in diverse presentations and lengths to affirm the dominant social group's rightful control of culture, religion, economics, and environment.

FIREWORKS AS CONTEMPORARY
NARRATIVE IN FICTION AND FILM

Fireworks have been widely used as a metaphor, whether or not the actual fireworks appear in the narrative. Their metaphorical uses can be seen in British author Angela Carter's first collection of short fictions, *Fireworks: Nine Profane Pieces* (1974), which uses postmodern narrative strategies (notably self-reflexivity) to interrogate mimesis and representation; some tales use actual fire*arms*, retaining the fireworks metaphor to allude to explosively oppressive social relations. Citing a narrative excerpt from the opening execution scene of the short tale, "The Executioner's Beautiful Daughter" ("this *tableau vivant* might be better termed a *nature morte* for the mirthless carnival before us is a celebration of death"), Julie Sauvage (2008) emphasizes the ways Carter "produces unreality effects in the narrative" through her extensive uses of tableaus, and her comparisons between theater (*tableau vivant,* a theatrical genre popular in the eighteenth century, using actors to stand motionless in imitation of famous paintings) and painting (still life, or *nature morte*) (127). In both of these genres, and by extension in Carter's *Fireworks* as well, nature is "killed" to produce art, a startling insight that Sauvage underscores by recalling theater's historical roots in animal sacrifice as a strategy for exculpating humans (i.e., *tragos* meaning the goat that is sacrificed as a substitute). In the aforementioned tale, the Executioner is a character tasked with punishing acts of incest with beheading, even carrying out this punishment on his own son—and wearing his Executioner's mask when perpetrating incest on his own daughter. Robbie Goh (1999) reads this tale as "a caricature of the modern police state" (76); it is also an apt reminder of the brutal rapes, murders, and incendiary deaths of women in Ciudad Juarez, a continuing violence which is perpetrated by drug rings of which the police themselves are members (Portillo 2001). But the connection between violent sexism, ageism, and the police state is only the start: through stories like "Master" and "Elegy for a Freelance," Carter illuminates the material conditions for women struggling to survive in the contexts of patriarchal colonialisms.

"Master" narrates a white hunter's colonial exploits in a South American jungle, where he abducts an indigenous girl for his sexual toy. Belonging to the jaguar clan, and "as virgin as the forest that had borne her" (*Fireworks*, 74), the girl's embodiment articulates the symbolic interchangeability of indigenous, female, youthful/immature, savage, and animal that is part of the colonialist Master Model which defines the "master" identity as adult, male, human, and European (Plumwood 1993). Plumwood's analysis offers a useful lens for understanding the ideology and practice that powered the European colonization and genocide of indigenous North America—including the literal rape of native women, as well as the violent assault on numerous animal species, destruction or pollution of forests, grasslands, and wetlands (Smith 2005). In "Master," Carter's enslaved indigenous female eventually shoots and kills the great white hunter with his own firearms. The final tale of the nine-fiction collection, "Elegy for a Freelance," focuses on "the lethal doubles of fireworks—firearms" (Sauvage 152) as a revolutionary prepares to commit an act of political terrorism, fighting back against an unjust nation-state. The tales in *Fireworks* illuminate the persistence of rape, and the linked and violent subordination of naturalized others via hierarchies of gender, species, age, race, nation, and ethnicity in colonialist conquest. Here as elsewhere, fireworks function as a sign that underscores rather than constructs specific cultural meanings.

In contrast, U.S. filmmaker Kevin Anger's film *Fireworks* (1947) is a fourteen-minute dream sequence that adapts and subverts themes from mainstream nationalism, using direct images of fire and fireworks to construct a gay fantasy of fulfillment. Completed when Anger was only seventeen, the film opens with the sleeping protagonist (played by Anger) awakening with an oversized erection (which turns out to be a wooden statue, when the sheet is pulled back) and going in search of the "light." He enters a door marked "Gents" and is assaulted by a gang of sailors in what appears to be a rite of initiation: fingers are shoved in the dreamer's nostrils and blood shoots out his nose and mouth; a sailor twists his arm and he screams; a bottle of cream is dashed on the floor, a broken piece thereof is used to cut the dreamer's chest, and cream poured from above flows over the cut and into his mouth (Kane 2008). The dreamer ultimately finds his "light" in "the sailor with a phallic roman candle who becomes a god-like figure, lying in bed with the dreamer at the film's end" (Meir 2003). Critics have noted the memorable image of the dreamer balancing a Christmas tree on his head, with a burning candle atop the tree: this tree/candle is used to set fire to photos of the dreamer in a sailor's arms. The film's conclusion pans from these photos burning in a fireplace to the bed, where the dreamer lies next to his male lover, whose face appears to be aflame (Anger scratched over the filmstrip to create the effect), then on to a plaster hand that had appeared with broken fingers at the start of the film, and now appears whole.

Anger's appropriation of mainstream nationalist images to describe his youthful gay perspective appears most notably in the translation of signifiers such as the muscular sailors, cream/milk, fireworks, and the Christmas tree. Instead of representing forces of U.S. Empire, the military is subverted through its representation as sailors who are the objects and agents of gay desire. The milk once signifying the nourishment of mainstream hetero-nuclear families (and rarely signifying its material origins in the anguished separation of cows from their nursing calves) is "a stand-in for seminal fluid" (Meir 2003), and Christmas trees are freed from signification confined to monotheistic and heteropatriarchal Christianity, becoming the evergreens of pagan ritual that translate the original solstice fertility into a homoerotic virility. Fireworks, of course, shift from the nation's patriotic Independence Day to a celebration of the "pyrotechnic phallus" that becomes the dreamer's celebration of gay liberation.

These readings of fireworks in cultural narratives suggest that fireworks are a metaphor, variously appropriated to underscore the hegemonic cultural messages that naturalize hierarchy, monarchy, male dominance, empire, class, nation, religion, ethnicity, species, and age. In narrative arts, fireworks have been used both to reinforce dominant narratives of monarchy, species, and sexuality and to subvert these and other hierarchies of age, empire, and gender (Angela Carter, Kevin Anger). Do fireworks "mean" anything in themselves, or are they just a mirror, a template on which society projects its own cultural images? On investigation, it appears the material features of fireworks—their chemistry; their effects on humans, animals, and environment; the labor involved in their production; their actual purchase and estimated social costs—matter as much as the narratives they convey.

A CRITICAL ECOFEMINIST PERSPECTIVE ON FIREWORKS

For centuries, fireworks created their visual, olfactory, and auditory effects from a combination of charcoal, sulfur, and potassium nitrate—gunpowder. Because these early blends of gunpowder were too unstable and messy, when the opportunity arose, the element that brings oxygen to the fire and speeds up the explosion, potassium nitrate, was replaced by perchlorates (potassium perchlorate and ammonium perchlorate). But perchlorates can remain in the air and water for days or weeks after a fireworks display, and pose significant risks for human, animal, and environmental health. The U. S. Environmental Protection Agency's Richard T. Wilkin and colleagues (2007) definitively established fireworks displays as a source of perchlorate contamination by analyzing water in an Oklahoma lake before and after fireworks displays in 2004, 2005, and 2006. Within fourteen hours after the fireworks, perchlorate levels rose 24 to 1,028 times above background levels. Levels peaked about twenty-four hours after the display, and then decreased to the prefireworks background within twenty to eighty days (Wilkin et al. 2007). Perchlorates have caused thyroid cancer in rats and mice, and have the potential to affect the thyroid gland in humans, resulting in hypothyroidism (McLendon 2009). The smoke from fireworks also contains other particulate matter that can get lodged in people's lungs, posing immediate danger to those with asthma or chemical sensitivities. And while perchlorates and other particulates dissipate within days or weeks after a fireworks display, the heavy metals that provide much of the coloration in fireworks persist.

To create their sparkling shower of colors, fireworks are loaded with heavy metals that include strontium, aluminum, copper, barium, rubidium, cadmium, mercury, lead, among others (McLendon 2009; Conway 2011). Barium produces the brilliant green colors and is used despite its known radioactivity, a property it shares with strontium (creating the red sparks), a metal that has caused birth defects in laboratory animals and is thought to impair bone growth in children. Copper (for the blue in fireworks) and perchlorates start the formation of dioxins, a well-known class of carcinogens that have also been shown to disrupt hormone production. Aluminum and cadmium both affect the lungs, along with their other distinct properties: aluminum is a suspected link to Alzheimer's disease, and cadmium is a known human carcinogen. Although many sources discuss the known properties of fireworks' heavy metals, they agree that the direct effects on humans are not yet known—despite numerous area-specific complaints of lung irritation, such as those made by neighbors breathing the fallout from fireworks displays at Disneyland in Anaheim, California.[3] Even a chemist at Los Alamos National Laboratory agrees that "everyone at or downwind of a pyrotechnic display is getting subjected to levels of [heavy] metals that aren't natural levels" (Sohn 2009).

Along with the animals used in laboratories to prove the toxicity of fireworks' heavy metals, companion animals, wild animals, and farmed animals are also harmed by pyrotechnics. According to the Ohio Animal Defense League, the Royal Society for the Prevention of Cruelty to Animals (RSPCA) of Australia, and numerous other animal welfare societies, a wealth of data confirm that fireworks frighten and confuse most animal species.[4] Animal shelters and humane societies regularly experience an up to 500% increase in the number of stray animals, reported injuries, and animal traumas after a fireworks event. Dogs have been brought to shelters with bloody paws from their attempts to outrun the explosions, or with torn skin from tearing through backyard fences, or worse, injured or crippled by a car collision. Animals who are too close to fireworks explosions can experience significant burns and eye damage, and the explosive sounds themselves are perceived at an intensity that is more acute than that perceived by human ears. A zoologist at the University of Guelph reported that the panic and disorientation birds experience during fireworks can result in the birds flying into a building, and researchers at Acadia University in Nova Scotia found that birds who nest in high densities are most at risk: after a loud bang, the birds will startle and fly away, becoming so disoriented that nesting mothers cannot find their own nests when they return. In 2011, one Alabama neighborhood's New Year's fireworks were set off near the roost of red-winged blackbirds and European starlings, producing booms so strong they shook the windows on houses, and startled these night-blind birds into a 25-mile-per-hour flight that dashed them into houses, signs, and the ground. Necropsies found trauma to the chest, hemorrhages, and the leading edge of the birds' wings broken (Weise 2011). Wildlife shelters experience an increase in orphaned birds, squirrels, raccoons, and other small mammals. Even farmed animals are affected: laying hens have very low egg production the day after a fireworks event, and the eggs they do lay are often malformed (Ohio Animal Defense League).

Worse than these effects are the malicious actions of humans who deliberately harm animals with fireworks. A vast international archive of these news stories includes headings such as "Cat found with horrendous burns" (England 2010), "Dublin thugs set firework off in dog's mouth" (Ireland 2010, 2009), "Sheep's jaw blown off with fireworks" (Mauao/New Zealand 2009), "Firework hooligans punished for horse's death" (Zwolle/Netherlands 2008), "Cat 'blown up' in firework horror" (Cyprus 2008), "Turtle firework cruelty condemned" (England 2007), "Fury as sick thugs throw fireworks at terrified cow" (Belfast 2007), "Rabbit blown up in firework horror" (Blackpool/England 2006), and "Fireworks injure zoo animals" (Malaysia 2004). What these and numerous other news stories confirm is the fact that fireworks are used as a tangible weapon of the warfare perpetrated on other animal species by specific humans—in the aforementioned cases, adolescent boys in westernized and industrialized nations.

Not only do fireworks pollute the air, water, and soil through their chemical residues and refuse, and offer cheap weapons of warfare to adolescents, another environmental justice concern about fireworks has to do with their manufacture by child slave laborers. The overwhelming majority of fireworks used in the United States, and perhaps in other countries as well, come from China, where they are manufactured by children between the ages of five and fourteen. China-watchers believe child labor is increasing, based on a high dropout rate from school in conjunction with the expansion of foreign investment in export-oriented enterprises (Grau 2005). Although China requires education up to age 16, in the economic zones of Guangdong, Sichuan, Zhejiang, and Hebei there are reported to be approximately four to five million child laborers under the age of sixteen, working ten to fourteen hours a day and earning half the wages of an adult. In March 2001, thirty-seven Chinese third- and fourth-graders were killed in a Chinese fireworks manufacturing-related explosion *at their school* (Llorca 2002). Despite these conditions, the United States imports pyrotechnics and explosives from China to celebrate its Independence, spending approximately $1 billion annually (Grau 2005). Child labor in fireworks production is also prominent in Guatemala and India, though the International Labor Office is working in Guatemala to eradicate child labor in fireworks production, and in India, there is a strong "Save the Childhood Movement" (Bachpan Bachao Andolan, or BBA) urging people to go fireworks-free for Diwali, the annual Hindu festival of lights. Like other child labor activists, they point to the manufacturing of matches and fireworks concentrated in Sivakasi and elsewhere in the state of Tamil Nadu, where a large number of accidents go unreported, and the constant exposure to chemicals like sulfur, potash, phosphorous, and chlorate cause the children to contract deadly infections of lung, skin, kidney, and eyes. Their fact sheet on the antifireworks campaign concludes, "Celebrate Diwali by lighting lamps, not by blasting childhood" (BBA 2010). Fortunately, in 2007, their work was augmented by a twenty-five-minute documentary film, *Tragedy Buried in Happiness*, shot by South Korean broadcaster Taegu Broadcasting Corporation with the help of Manitham, a human rights NGO working with children, Amnesty International, and the National Confederation of Human Rights (Menon 2007). Unable to persuade an Indian NGO or filmmaker to shoot the film, Manitham's executive director turned to Korean filmmakers (the film is shot in Korean, and dubbed into Tamil and English). Their film profiles children whose work in fireworks manufacturing has scarred their faces and hands, and who still must work from 7:00 a.m. to 6:00 p.m. for between INR 40 and INR 100 a week (the equivalent of US$1.98). Their labor conditions exemplify the "slow violence" of global environmental racism (Nixon 2011).

Over the past century, fireworks violence has affected not only "scapegoated" animals and third world enslaved children, but also children and adults in the industrialized world. An early record kept by the Chicago *Tribune* detailing July 4 injuries dating from 1899 to 1908 lists 508 deaths and 29,085 casualties ("A Quiet Fourth" 1909). In contrast, the American

Medical Association's (AMA) record for the 6 years 1903–1908 lists 1,316 deaths and 27,980 wounded, largely because the *Tribune* recorded only July 4 incidents, whereas the AMA recorded all fireworks-related injuries. In 1909, 215 people were killed, many of them children ("In our Century" 2010), and in 1913, the *Journal of Education* reported 32 deaths, 13 of these from little girls who were burned to death when their clothing caught on fire ("Fourth of July Casualties"). Given the nation's respectively smaller population size in the early 1900s, these numbers are significant when compared with the 2010 Annual Fireworks Report from the U.S. Consumer Product Safety Commission, which found only 3 fireworks-associated deaths in 2010, but an estimated 8,600 fireworks-associated injuries treated in emergency rooms (Tu & Granados 2011).

Meanwhile, U.S. consumers flock to "the most extravagant fireworks displays on July 4" (Berr 2010), which occur in Philadelphia (expenditures reduced from $3 million to $2.1 million), Boston ($2.5 million), Seattle ($500,000), Pasadena's Rose Bowl ($350,000 to $400,000), and Addison, Texas ($240,000). Notably, the sponsors of New York City's fireworks (Macy's) and of Washington, DC, "declined to comment" when asked to divulge the costs of their Independence Day fireworks extravaganzas. Admittedly, smaller towns, such as Great Falls, Montana, produce smaller fireworks celebrations—$15,000 in 2011 (City of Great Falls)—but the funding of fireworks effectively de-funds other public projects. Many social and environmental justice activists question the value of these ephemeral celebrations in place of more lasting investments in community and national well-being, such as updating the drinking water-delivery infrastructure in the United States, rebuilding bridges and highways, funding public education and public libraries, providing greater access to health care, increasing affordable housing, enhancing mass transit systems, maintaining public parks, and protecting wildlife. More lasting investments in global well-being could be made by withdrawing the funds used to purchase fireworks in first world nations and channeling these funds to third world NGOs working to build local ecological and economic sustainability, educate children, and protect girls and boys from sexual and economic slavery. It seems that on both ends of fireworks production and purchase, significant material costs are borne by those least able to afford them—third world children, poor and working class Southerners, animals, the environment.

CONCLUSION

In 2008, women in Naples, Italy, organized a classic protest to stop fireworks celebrations that injure or maim hundreds each year. "Se Spari, Niente Sesso"

(If you shoot, no sex) was inspired by Carolina Staiano, a mother of two, whose father became partially paralyzed in a fireworks accident ("Women use threat" 2008). As in Aristophanes' Greek comedy, *Lysistrata*, wherein women refuse to have sex with men until the men stop creating warfare, these contemporary women organized a group that withholds the erotic from those whose fireworks "celebrations" actually harm the eros of life itself.

The militarist/celebratory, firearms/fireworks dualism underlies these women's protests. As Plumwood's Master Model shows, dualisms are produced by separating concepts that are at root inseparable. Fireworks' death-defying dominance over nature and firearms' death-dealing dominance over the Other are more linked than separate, though consumers of fireworks' spectacles would rather keep them apart, focusing only on the elite-approved, celebratory, and sublime aspects. A critical ecofeminist perspective traces the branches of firearms and fireworks down to their shared root: down through the environmental injustices of the death-dealing slow violence inherent in child slavery, workplace injuries, and economic injustice; down through the material facts of environmental toxins, through the human-animal studies' recognition of multiple species' injuries and deaths, down to the root of multiple and linked toxic narratives celebrating hyperseparation, colonialism, warfare and dominant masculinities. Certainly, fireworks displays are breath-taking; otherwise their narratives wouldn't be as effective, and spectators might look beyond the sparkling lights into the penumbra, where they would see the enslaved children, the fleeing and terrified birds, the brutal rapes and killings of colonial warfare that in the United States enabled these independence celebrations on stolen land.

Instead of identifying with the Master side of the Master/slave dualism, a critical ecofeminist perspective restores connections, shifting the focus from the approved stories of centralized elites to the narratives from the margins. This shifted focus reconstructs the subject, from the nationalist stance of spectator-citizenship to a globally conscious stance of engaged interidentity. Without allegiance to a nation-state, the social construction of an enemy-other falls away, and militarism might be displaced by connection, sustainability, and economic justice.

Re-stor(y)ing the material narrative of fireworks' "truncated narrative" (Kheel 1993) requires writing new stories that illuminate the ways fireworks have been a "fun" celebration for the few, at the expense of the many. In Susan Meddaugh's "Martha Speaks" series about a talking dog, *Fireworks for All* (2011) presents Martha organizing her canine companions to gather signatures for a fireworks ban—despite the fact that her human family ignores Martha's information that "fireworks are scary to dogs" and despite the fact that the woman collecting signatures for the ban doesn't like dogs! Only after the ban is passed does Helen (Martha's human girl companion) acknowledge

Martha's interests and propose a solution that pleases children and dogs alike—offering a special evening show of "Courageous Collie Carlo" at the movie theater, just for dogs, who will thus be protected from hearing the fireworks. Admittedly, this children's book does not address the range of other species, workers, spectators, and environments affected by fireworks, but its narrative offers an opening for others to follow.

More ecologically just solutions involve finding alternative ways to celebrate, such as the Vatican's "Endangered Species Light Show" scheduled in November 2015 to coincide with the COP21 Paris Climate Change Summit: for an hour, fifty projectors showed images of endangered species, illuminated on the "canvas" of St. Peter's Basilica in Vatican City.[5] Environmentalists, animal advocates, and labor activists alike suggest organizing laser shows, community drum circles, block parties, or star-gazing celebrations in place of fireworks. There are also "green" fireworks still under development, ones that burn nitrogen-based fuels rather than using perchlorates and emitting ten times less barium (Sohn 2009); however, these are still too expensive for neighborhood shows, and they fail to eliminate the other forms of pollution—noise, light, and (shall we venture to say) ideological—that are a mainstay of fireworks' history.

What's at stake in these diverse cultures' spectacular narrative of fireworks is the reinforcement of a dominant group's *naturalization* of empire, erotophobia, and ecophobia—narratives whose slow violence culminates in apocalypse.[6] For the well-being of this precious earth and its many inhabitants, we need new celebrations of interdependence, and new narratives of celebration.

NOTES

1. The "Star-Spangled Banner" lyrics were written by Francis Scott Key in 1814 and set to a pre-existing melody. The song has been played or sung at Independence Day events as well as sports games, high school graduations, military, and civil ceremonies of many types for almost two centuries. It became the U.S. national anthem in 1931. Complete lyrics are available on the USA Flag Website (http://www.usa-flag-site.org/song-lyrics/star-spangled-banner.shtml).

2. Guy Fawkes was in charge of the explosives that were to be used in the Gunpowder Plot of 1605, a plan to assassinate the Protestant King James I and restore a Catholic monarch to the throne. Fawkes was discovered early in the morning of November 5, and he was captured, tortured, and interrogated. Sentenced to be executed, Fawkes allegedly broke his neck by jumping off the platform where he was to be hanged and later drawn and quartered, causing his own death and avoiding much additional suffering. His effigy was burned on a bonfire, re-enacting the public punishment of Catholics who would usurp the British throne. Guy Fawkes Day was soon written into law and become the predominant English state celebration, or

nationalism with anti-Catholic overtones. It is thus seen as more of a British holiday than an Irish one.

3. Disneyland has switched to air-propelled fireworks, eliminating the perchlorates but not affecting the aspects of noise and light pollution.

4. Australia's Royal Society for the Prevention of Cruelty to Animals (RSPCA) was inspired by the RSPCA in the United Kingdom. The Ohio Animal Defense League offers an impressive array of statements about fireworks and their effects on domesticated and wild animals from animal welfare organizations around the United States; see http://www.all-creatures.org/oadl/quot.html

5. See https://www.youtube.com/watch?v=z4G5o4oqEtk for a half-hour clip.

6. For the concepts of "erotophobia," "ecophobia," and "slow violence," see respectively Gaard (1997a), Estok (2009), and Nixon (2011).

Chapter 5

Animals in Space

A term biologist Eugene F. Stoermer invented, but Dutch chemist and Nobel Prize-winner Paul Crutzen popularized, "Anthropocene," refers to an era of human-induced atmospheric warming that can be traced back to the Industrial Revolution, when specific communities of humans began increasing carbon dioxide emissions by burning coal and oil, building larger and larger cities, cutting down forests, acidifying oceans, and prompting massive species extinctions (Revkin 2011). Despite numerous warnings from scientists and scientific organizations around the world, responses to climate change among the most industrialized nations have been slow to put long-term ecological sustainability and health ahead of short-term corporate profits. Instead, like Bill Peet's children's book, *The Wump World* (1970), global elites, politicians, and business leaders are behaving like The Pollutians, assuming that they can travel from continent to continent, and planet to planet, polluting and then moving on to new pristine environments without changing their ecological and economic behaviors. Ideologically fueled by literal interpretations of transcendent theologies which locate heaven and the sacred in the skies above earth, and by Western techno-science's quest to control nature, space programs in both the United States and Russia have appropriated public monies to fund their search for otherworldly escapes available only to earth's elites, diverting public funds that would otherwise be used to meet real material needs (housing, healthcare, education, food security) or fund research and infrastructure for promoting sustainable energy production, transportation, and agriculture on Earth, the only human-habitable planet in our solar system. As Andreé Collard and Joyce Contrucci once wrote in *Rape of the Wild* (1989),

> There is nothing transcendent about the values that motivated NASA to have astronauts lodge an identification plaque on the moon indicating that men had landed there. This gesture grates as much as coming upon a tree or a rock defaced with "John loves Mary" or some similar nonsense. . . . There is really no difference between the "humanization" of space and the colonization of Africa or Latin America. (1989, 166)

What can a critical ecofeminist perspective contribute to our understanding of space exploration ideology? How do narratives of gender, species, class and culture play out beyond the biosphere? And what information do these scientific pursuits of the twentieth century have to tell us about contemporary environmental problems and solutions for the future?

To explore these questions, I juxtapose three parallel narratives testing the limits of extra-terrestrial exploration and survival: the use of non-human animals in space exploration as a precedent to sending humans into outer space; the conception, missions, and ultimate failure of Biosphere 2, a facility constructed to replicate five of Earth's biomes, and test the possibility of indefinite human survival within a sealed enclosure; and the current ventures of NewSpace corporations in cultivating space tourism and settlements. Drawing on feminist philosophy of science, feminist animal studies, and ecofeminist theories, this chapter develops a critical ecofeminist perspective located at the intersections of gender, species, race, class, and culture in space exploration narratives.

"UNDER MY THUMB": CRASH TEST DUMMIES AND ONE SMALL STEP FOR (A) MAN

Notwithstanding arguments that space travel has produced greater environmental awareness via Apollo 8 images of *Earthrise from the Moon* and Apollo 17's image of the *Whole Earth* (Henry & Taylor 2009), in this section, I argue that space exploration is advanced within a framework of masculinist ideology that values a type of holism over specific individuals, heroic feats of conquest amid risk-riddled adventure, and techno-scientific solutions to the ecosocial problems produced by runaway capitalist imperialisms now warming the earth. The astronauts who suddenly discover a deeper respect for the earth when they are able to blot it out with their thumbs[1]—a gesture of dominance reminiscent of Mick Jaggar's "Under My Thumb," a 1966 lyrical celebration of dominance over his "squirming dog," "siamese cat" girlfriend-now-turned-pet—do not offer a pathway to environmentalism that can (or should) be widely duplicated, even via the pervasive Whole Earth images commodified on calendars, key chains, and coffee cups. Instead, it is the particular relations of animals, places, and cultures that require our environmental and climate justice concerns, bringing us to the roots of contemporary ecojustice crises rather than striving for ever-greater techno-science explorations of space.

Feminist philosophers of science have amply noted the gendered features of the "scientific method" which requires a cutting-off of feelings to produce the "detached eye of objective science" and the distancing of the scientific researcher from the experimental subjects (Haraway 1989, 13; Keller 1985; Keller & Longino, 1996). Donna Haraway's *Primate Visions* demonstrates that Western science's construction of the scientific standpoint is inflected by race, gender, and species supremacy, controlling not only scientific rhetoric and investigations, but also Western culture's relationship with nature and other animal species.[2] In *Nature Ethics: An Ecofeminist Perspective,* Marti Kheel identifies features of masculinism that are not only inflected with race and species supremacy, but also embedded in concepts of rationality, universality, and autonomy (2008). These features are also evident in the rhetoric justifying space exploration, including the belief that humans (particularly those gendered masculine) are propelled by aggressive, self-centered biological drives that must be given controlled, rational expression; the (racist and imperialist) idea that nations must preserve the "frontier" experience as a legacy for future generations, especially boys; and the notion of adventure (especially high-risk) as counter to and not found in the repetitive realm of biological nature. As Kheel explains, masculinism is inherently antiecological for the ways it "idealizes transcending the biological realm, as represented by other-than-human animals and affiliative ties," and subordinates "empathy and care for individual beings to a larger cognitive perspective or 'whole'" (2008, 3). Across the disciplines, men's movement writers, animal studies and environmental studies scholars expand Kheel's critique, identifying numerous constructs of masculinity as predicated on themes of maturity-as-separation, with male self-identity and self-esteem based on dominance, conquest, affects (work ethic and emotional stoicism), occupations (valuing career over family and housework), physical strength, sexual prowess, animal "meat" hunting and/or eating, and competitiveness—all developed in opposition to a complementary and distorted role for women: white heterohuman-femininity (Adams 1990; Connell 1995; Plumwood 1993). As the conflicted histories of chimponauts and astrodogs demonstrate, narratives of space exploration are constructed within this larger narrative of masculinist gender ideology that has shaped definitions and practices of science itself.

More than a decade before the U.S. National Aeronautics Space Administration (NASA) produced Neil Armstrong's famous moonwalk on July 20, 1969, American and Russian scientists used nonhuman animals—mostly monkeys, chimpanzees, and dogs—to test the effects of rapid acceleration, prolonged weightlessness, atmospheric reentry, and other hazards of space travel. To obtain these animals, the United States funded the capture of young and infant chimpanzees from Africa for space exploration tests at the Holloman Air Force Base in Alamogordo, New Mexico; some sources say the chimpanzee mothers were killed in order for their babies to be taken (Cassidy & Davy 1989). In Russia, Soviet scientists took stray dogs off the streets of Moscow. In the cultural ideologies of both nations—intensified by the Cold War—space colonization became a matter of nationalist pride, and the "sacrifice" of nonhuman animals was seen as a necessary precedent to "manned" flights.[3]

Beginning June 11, 1948, when the first mammal in space, a Rhesus monkey named Albert I was launched at the U.S. White Sands Proving Ground in New Mexico, a series of Rhesus macaques named Albert 1 through VI were launched and either died on return impact or died from heat prostration following recovery. In 1959, the first U.S. monkeys to survive, Able (a rhesus macaque) and Baker (a squirrel monkey), were soon followed by flights with chimpanzees, Ham (January 31) and Enos (November 29) 1961. Originally nicknamed "Chop Chop Chang," chimpanzee #65 wasn't given his official name—an acronym dubbed after the *H*olloman *A*ero-*M*edical Research Laboratory where the space chimps program developed—until it was clear he had survived his seventeen-minute flight. His name, Ham, also "inevitably recalls Noah's youngest and only black son," exemplifying the "stunning racism" in the language of the space program, constructing the paternalistic identity of the scientist as well as the scientific endeavor (Haraway 137–38).

While space race fans have claimed that Ham's flight was a necessary precedent to the first "manned" U.S. suborbital flight of Alan Shepard in 1961, and Enos' over three hours and two-orbits demonstration was a precedent for John Glenn's first U.S. orbital flight in 1962, these heroic human volunteers were simply following in the involuntary handprints of their chimpanzee predecessors. Indeed, one wonders how the space race would have proceeded if nonhuman animals were not available as "crash test dummies," and each test flight would have had to be piloted by computer, by a human model, or by a sacrificial and highly trained human volunteer.[4] The fact that the animal lives used and often destroyed in space exploration were treated with some indifference can be read in the documents describing their deaths: the sacrificial rhesus monkey Albert VI was nicknamed "Yorick" (alluding to Shakespeare's Hamlet and his graveyard soliloquy with the court jester's skull[5]) and a U.S. Air Force photo shows a small memorial, complete with three rubber mice, plastic flowers and a flower vase, with a card that reads "Sincerest condolences to Thee, our departed friends of Discoverer III, from the Army Monkey" (Burgess and Dubbs 186). Testing the effects of high gravity forces ("g-forces"), acceleration and rapid deceleration that might occur on rocket flights, chimpanzees, bears, and hogs were strapped in various positions (sitting up or lying down, head-first, facing forward or back), with and without safety harnesses, on sleds titled "Gee-Whizz," "Sonic Wind," "Project Whoosh" and the "Daisy Track," whimsical names constructed from the standpoint of those who did not ride the tracks at lethal rates. One hog photographed in a crash simulation harness (sitting up, facing backward) albeit with the head lolling to one side (it's not clear whether the animal photographed is dead or alive, before or after the test) has a sign resting beneath the beltstrap and between the legs, reading "Project Barbecue, Run #22, 5 August 1952" referencing the fact that after the hogs had suffered, died, and been autopsied, they were cooked and eaten by the Air Force scientists (Burgess and Dubbs 105). The rhetoric of these aeronautics scientists' treatment of animals reinforces the dominance, the adventure, and the unfeeling identity of the scientists and the scientific project at hand.

While the United States was experimenting with monkeys, the Soviet Union was experimenting with dogs. Scientists preferred the small female strays taken from the streets of Moscow since females needed less room to urinate, and could be more easily trained for space flight. Upon capture, the dogs were confined in small places, subjected to extremely loud noises and vibrations, and made to wear newly created space suits, all tests designed to condition the dogs to the experiences they would likely have during the flight. The first dogs launched, Moscow's Tsygan and Dezik, reached space on July 22, 1951, but did not orbit; however, they were the first mammals successfully recovered from spaceflight. In the next few years, Russia launched numerous dogs into suborbital flight with at least four fatalities, but Soviet scientists were eager to make some "sacrifices" in the Cold War race to beat the United States in moving toward outer-atmosphere orbits (Kemp 2007). At the request of Premier Nikita Khrushchev and the orders of Chief Engineer Sergei Pavlovitch Korolev, on November 3, 1957, to celebrate the fortieth anniversary of the Russian Revolution, a thirteen-pound, three-year-old female stray dog was launched into Earth orbit in Sputnik 2. Originally named Kudryavka ("little curly") and later renamed Laika ("Barker"), the little Samoyed-husky dog had been selected for her obedience and calm disposition. Sputnik 2 was an impromptu mission built only a month after the internationally acclaimed success of Sputnik 1, leaving the Soviet engineers no time to design provisions for the dog's return from space.

Evidence that the Soviet scientists were conflicted about their duties is recorded in events leading up to Laika's launch, when one of the dog's trainers, Vladimir Yazdovsky, took Laika home to play with his children; later, he wrote in his own account of the mission, "I wanted to do something nice for her. She had so little time left to live" (Oulette 2011). A full three days prior to launch, Laika was strapped into the space capsule on October 31,

1957, to monitor her vital systems. On launch day, November 3, Yazdovsky and the medical staff persuaded the engineers that Laika's capsule must be de-pressurized—and then used this change as an opportunity to give Laika her last drink of water (Burgess and Dubbs 159). Their actions suggest an emotional turmoil produced by the conflict of "entangled empathy" repressed under obedience to Cold War nationalism and the cultural constructions of masculinized science (Gruen 2012).

But the international viewing public was less obedient. As soon as Laika's launch aboard Sputnik 2 was announced to the press, animal-welfare groups around the world expressed outrage and sorrow: in Britain, protesters assembled at the Russian embassy, and the National Canine Defense League called for a minute of silence each day that Laika was presumed to be in orbit. The initially deceptive Russian news releases soon had to acknowledge that there were no plans for Laika's return to Earth, and though they suggested she remained healthy for several days, over forty years later, Dimitri Malashenkov from the Institute for Biological Problems in Moscow finally admitted that Laika became stressed (her heart rate accelerated to three times its normal rate) and overheated, most likely dying a painful and terrifying death. In 1998, Oleg Georgivitch Gazenko, one of the Soviet scientists responsible for the dogs' training, admitted "the more time passes, the more I'm sorry about it. We did not learn enough from the mission to justify the death of the dog" (Oulette 2011).

In the immediate wake of Sputnik 2, Soviet nationalists attempted to construct Laika's capture, confinement and death as an act of heroism: photographs issued by the space agency, with Laika exuding "an air of bright courage," were used on Mongolian and Romanian postage stamps and souvenirs. A Monument to the Conquerors of Space was built in Moscow and inaugurated on October 4, 1964, featuring Laika's turned head and a trace of the space harness (Kemp 2007).[6] Even the U.S. space program was not indifferent, with NASA naming a soil target on Mars after Laika. But on both sides of the space race, the nationalist gratitude for animal lives lost in space exploration has been oddly expressed: taxidermists have stuffed Strelka and Belka, the first animals to orbit Earth and return alive, and they are now on display in the Memorial Museum of Astronautics in Moscow; Ham's remains are buried at the entrance to the International Space Hall of Fame in New Mexico. And after the space race ended, the U.S. Air Force began leasing the remaining chimpanzees at Holloman Air Force Base to medical labs in the 1970s, and in 1997 "retired" the space chimps to a biomedical testing facility, the Coulston Foundation, which had a known and horrific track record of abusing chimps. Over the years, USDA investigations had found Coulston in violation of numerous animal welfare codes, and at one point confiscated 300 of their chimps. Finally, Dr. Carole Noon, with the backing of Drs. Jane Goodall and Roger Fouts, worked to bring these chimpanzees to sanctuary.[7] These postmortem heroic narratives fail to conceal the speciesism and animal suffering produced under the name of science.

Illustrating many people's discomfort with the treatment of Ham, Enos, and Laika, recent retellings of these stories in children's literature and media have attempted to explore and make palatable this anguished past.[8] In 2007— the fifty-year anniversary of Laika's "one-way" flight— James Vining's *First in Space* appeared, a graphic novel detailing the life of Ham, along with two other children's books about Laika using the dog's own viewpoint as part of the narrative: both Nick Abadzis' *Laika* and Jan Milsapps' *Screwed Pooch* detail the historical events of the Cold War and the space race that led up to the capture, training, and selection of Laika as the first and (allegedly) only creature knowingly sent into space to die. As Abadzis' novel clearly portrays, the founder of the Soviet Space Program, Sergei Pavlovich Korolev agreed to sending a dog in Sputnik 2 in order to prove his patriotism to Premier Khrushchev: after spending nearly eight years in Stalin's concentration camps under false allegations of sabotage, Korolev had worked his way up in the Soviet Space Program and gained respect for his energy, intelligence, and ambition, but was not yet pardoned for the false charges against him. In a particularly insightful scene, Abadzis draws Korolev into the space dogs' caged enclosure for a soliloquy with Laika, where Korolev reflects on his own imprisonment, and recognizes the parallels with Laika's confinement, but concludes that he cannot set her free: still on parole, Korolev believes his freedom might be achieved through her death. In return, he promises to make her "the most famous dog in history"—a reputation that matters very little to dogs (Abadzis 2007, 132). The scene is immediately followed by and contrasted with Abadzis' fictional character, Yelena Dubrovsky, the dogs' trainer, who articulates the story's ethical concerns: urging Korolev to choose a different dog, she offers to do "anything," an innuendo that is not lost on him. As a woman being used in a male-dominated system of masculinist science, Yelena exhibits the conflict between nationalist loyalty and entangled empathy, which Lori Gruen (2011) defines as an empathy that "requires gaining wisdom and perspective and, importantly, motivates the empathizer to act ethically." Through the fictional character of Yelena, and through his thematic treatment of the cycles of abuse that converge in Laika's death, Abadzis opens the dog's story for a critical ecofeminist reading of space exploration science and its cultural ideology.

Half a decade after these events, it's easier to see how the real lives of these specific, individual animals—their capture, confinement, training, and deaths—were backgrounded[9] by the dazzling material and discursive rhetoric of space exploration, as even the coauthors of *Animals in Space* recall:

One November night in 1957 our rowdy [Boy Scout] cub pack had been herded out of the scout hall at a certain time and made to stand under the crystal clear night sky while our cubmaster patiently told us about Laika and Sputnik 2. Suddenly he pointed with excitement above the darkened horizon, and we quickly fell into an awed silence as we watched a small, bright pin-prick of light silently and majestically traverse the star-spangled firmament over the east coast of Australia. (Burgess, xvii–xviii)

I cannot overstate how indelibly the image of a dog in a satellite burned into my youthful imagination. For me, at the age of 11, there was simply no way to comprehend it. It was too novel, too extraordinary an achievement, that it did not fit within any knowledge base that I possessed. It was mythic. . . . *I marveled more for the extraordinary experience given to Laika than I agonized over her fate.* (Dubbs, xix; italics mine)

Like fireworks, the space capsules streaming through the skies offered a visual and material narrative of celebration and heroism especially suited to Euro-Western constructions of dominant masculinity—with images that appealed to little boys and Air Force scientists alike. From the chimps' diets of baby cereal and baby diaper clothing to the "team of tender technicians" who put Enos into "a fitted contour couch that looked like a cradle trimmed with electronics," the visual contrasts between tall, white laboratory-coated human men and the small, diapered, and telemetrically implanted young mammals reinforced the masculinity of Cold War science and the "Father Knows Best" authoritative stance of white patriarchs from the United States to Russia.[10] As one of Laika's trainers, Oleg Georgivitch Gazenko, acknowledged forty years after her death: "We treat them like babies who cannot speak" (Abadzis 2007, 201).

Experiments involving other animals' bodies and lives to obtain information of primary interest to humans have long ago been exposed through well-developed critiques in animal rights theories of the 1970s and animal ecofeminisms of the 1980s as experiments that are often repetitive, painful, frightening, and unnecessary, given the less-expensive nonanimal alternatives; two decades later, such acknowledgements are finally appearing in academic theory as well.[11] Powering and legitimating such scientific experimentation are certain beliefs about what "counts" as scientific research methods, methodologies, and epistemologies (Harding 1987). Feminist approaches to science differ from traditional (androcentric) science not merely by "adding" women to science, whether as researchers or as subjects worthy of study, but in the ways feminists approach these core beliefs. Feminist methodology requires praxis, an activist approach to scientific research that seeks information to increase understanding and improve real material conditions for marginalized individuals and communities, particularly those under study. Thus, feminist methods require "listening carefully" to women and other marginalized beings for the data provided through their experiences and perspectives, and listening "critically" to how traditional scientists describe this data, seeking out information that traditional scientists "have not thought significant" (Harding 1987, 2). Finally, feminist scientists ask questions about what counts as knowledge, who can be a "knower" or "agent of knowledge," and effectively reconstruct the very identity of the scientist: rejecting the detached, authoritative "context stripping" objectivity of rationalist science, and its reason/emotion value dualism, feminists emphasize the inseparability of subjectivity and objectivity, locating the researcher on the same critical plane as the subject and cultivating the "authority" of both standpoints in the research project (Keller 1985; Hubbard 1990; Birke 1995; Mayberry, Subramaniam & Weasel 2001). Feminists regard as research assets the fundamentally relational character of human intersubjectivity, reason, and emotions.

Ecofeminists have long ago rejected the highly gendered reason/emotion dualism and the elevation of groups over individuals that characterizes not only Peter Singer's (1975) utilitarian ethics, but also the environmental ethics of "holism" that subordinates empathy and care for individual beings to a larger cognitive perspective or "whole" (Kheel 2008). Instead, ecofeminists and feminist animal studies scholars base ethics on the feelings and reasons that emerge from our relational interidentities, using the language of care, compassion, sympathy, and empathy (Adams and Donovan 2007; Gruen 2014; Donovan 1990, 2014). They note the linkages among diverse systems of oppression, whether these be the abuse of women, children, and nonhuman animals; among racism, sexism, and speciesism; or among the oppression of indigenous people, nonheterosexual behaviors, and nature (Adams 1995; Gaard 1997; Garbarino 2007; Harper 2010; Kemmerer 2011). Finally, they reject the elisions of evasive language that background the suffering and death of nonhuman animals (i.e., "veal"), and the emotionally distancing language of traditional science's particular brand of humanism that can be seen in the terms of "payload" for the living animal trapped aboard a space shuttle; "Chop Chop Chang" and "#65" for a chimpanzee who might not survive the space flight; "sacrifice" for the capture, confinement, training, vivisection and deaths of animals used in science; and Laika's "fate" or "destiny," as if her death was something inherent in her being, and not something produced through the agency of the Soviet space scientists.

The antifeminist, antiecological characteristics of space exploration as it has been practiced are amply evident in the economics, methodology, and ethics of the space programs described here. Post-World War II funding for space exploration diverted government funds away from other public projects,[12] all the while arguing that the benefits of space exploration would apply to all of humanity. But the Cold War space race between Russia and the United States belied those claims, suggesting that masculinism, nationalism, and colonialism were stronger motivations than humanitarianism.

BIOSPHERE II: ESCAPE TO INNER SPACE

Two vast and trunkless legs of stone
Stand in the desert. Near them, on the sand,
Half sunk, a shattered visage lies, whose frown,
And wrinkled lip, and sneer of cold command,
Tell that its sculptor well those passions read
Which yet survive, stamped on these lifeless things . . .

—Percy Bysshe Shelley, "Ozymandias" (1818)

To link Russia's "cosmodogs" and NASA's "chimponauts" with Biosphere II, one needs to consider not only the themes of space exploration and the colonizing drive to be "first"—first in space, first to orbit the earth, first to send a man into space, first on the moon, first woman in space—but also the question of enclosing animals in space missions. None of the dogs or chimpanzees sent into space cared anything about "firsts" or fame; nor did the animals confined in Biosphere II to nourish the human animals also confined there care or benefit from the multimillion dollar experiment going on in Oracle, Arizona. Renowned primatologist Jane Goodall has spoken to both ventures, explaining that Ham's apparent grin of happiness upon his return to Earth actually signified "the most extreme fear" through his baring of teeth, and admonishing the Biosphereans midway in their two-year enclosure that their confinement was far less than that experienced by caged chimpanzees, who "do not have the mental capacity to understand what is occurring or how to deal with it" (Cassidy and Davy 1989). But when Goodall returned almost a year later to deliver the final remarks concluding the Biosphereans' two-year enclosure (and extended their mission by twenty minutes) at least one Biospherean fumed, "Jane, let us apes out of the cage!" (Poynter 2006, vii). Goodall's instruction in the differences between voluntary and involuntary confinement had been lost on the humans, but the parallels of confining animal bodies to serve human cultural conceptions of masculinized astroscience are well worth exploring.

Envisioned in continuity with the space shuttle missions of the 1960s, Biosphere II was built with the two-pronged intention of developing an earth-based shelter for humans—anticipating an uninhabitable future on earth—and providing "the first model and the data . . . that will allow the successful building and operation of the Mars settlement" (Allen 1991, 75). The charismatic leader of the Synergia Ranch community and visionary for Biosphere II, John Allen envisioned biospheres as "refuges for a small elite from nuclear war or other disasters," believing "higher forms of life" could survive on "their own energy resources in mountain caverns" and "release full-scale life" back to Earth "after the skies began to clear" (Broad 1991).

In Allen's book *Space Biospheres* written with "Biospherian" Mark Nelson, they explain that "the major motivation behind creating Biosphere 2 . . . is to assist the Biosphere [meaning, 'Biosphere I,' our global ecosystem] to *evolve off planet earth* into potential life regions of our solar system" (Allen and Nelson 1989, 3). The metaphors describing Biosphere II tended to naturalize the project—that is, "Spaceship Earth" becoming "Biosphere I" and Biosphere II becoming another spaceship like Earth—and reveal the hubris of its creators. Roy Walford, the doctor involved in the project, called it "the Garden of Eden above an aircraft carrier" and *Time* magazine even called it "Noah's Ark: The Sequel" both metaphors referencing the grandiosity of divine creatorship assumed by both Allen and his followers (Smith 2010). Thus, despite any professions to the contrary, Biosphere II was a deeply antiecological project: instead of seeking ways to nourish living ecosystems by balancing human populations, consumption and waste behaviors, and challenging the economic and political forces affecting those ecosystems, Biosphere II exemplified the "truncated narrative" (Kheel 1993) obtained from the conjunction of heroic masculinist ideology, technology, and the environmental *science*s, operating in a neoliberal framework without the benefit of knowledge and perspective from the environmental humanities—that is, environmental economics, environmental ethics, critical animal studies, environmental justice, climate justice, food justice, and ecopsychology. Biosphere II offered "a glimpse of where 'sustainable development' might lead," wrote Timothy Luke, "if 'sustainability' is viewed as a purely technical and managerial problem" (Luke 1995, 159).

From the start, the project's vision was powered not by science but by ideology and money—namely, Ed Bass, a Texas billionaire and heir to an oil and real estate fortune, who eventually funded the project with $200 million, more than any governmental agency could afford. In 1984, he formed Space Biospheres Venture (SBV) with Margret Augustine and John Allen, the charismatic leader who had already founded the Institute of Ecotechnics, inspired by 1960's values of communal living, meditation, theater, and Buckminster Fuller's concept of synergy and his view of "Spaceship Earth." Housed above an art gallery in London, the Institute of Ecotechnics (IE) became involved with projects around the world—not just Synergia Ranch near Santa Fe, New Mexico, but an ocean-going "research" vessel the *Heraclitus*, a cattle station in the Australian outback (Quanbun Downs), and the Caravan of Dreams, a performing arts center in downtown Fort Worth. Eventually, the Institute of Ecotechnics granted "degrees" to many of the Biosphere II staff, who otherwise had no college education; for example, the "co-architect" for Biosphere II, Margret Augustine, was discovered to have no architectural training at all, apart from the IE diploma (Siano 1992, 41). Part of their theater training in the Caravan of Dreams, the Biosphereans were well aware of the theater of Biosphere II. John Allen's coauthor and staunch Synergist, Mark Nelson, reportedly told a colleague, "We do whatever we need to do, and play what roles we need to play, to get done what we need to get done"; according to Jane Poynter, this approach is "liberating" because "one does not get hung up on how credentialed a person is, but instead focuses on how competent he or she is in the role" (Poynter 2006, 237). In Biosphere II, Synergists played roles as "captain, head of the Agriculture, doctor, or analytical chemist" performed a NASA-like theater, complete with terms like "launch date" and "Mission Control," and costumes of coral-red space jumpsuits with matching boots (Cooper 1991a).

To prepare—and qualify—for their journey inside Biosphere II, a number of people associated with the assembled Synergia community underwent journeys on the *Heraclitus*, spent time in the Australian outback at the cattle ranch, and otherwise lived as invited (or directed) by the leadership team of John Allen and Margret Augustine.[13] These eager contestants gathered token species from across the globe, all selected for their usefulness to human life and energy conversion processes, to create their Noah's Ark of six biomes: a tropical rainforest, an ocean with artificially generated waves and a sub-real coral reef, a marsh estuary bridging the ocean and a fresh-water pond, a savannah with plants from three different continents, a desert with plants from four continents, and an agricultural zone which included both plants and animals—fish, goats, pigs, and chickens. The human zone contained both public and private spaces, a library and a kitchen above ground, with the "technosphere" below ground, where all the motors and the "lungs" of the system operated.

Ostensibly intended as a two-year project testing the viability of a self-contained system that recycled air and wastes alike, Biosphere II quickly ran into barriers that were both material and scientific, as well as social and psychological: after just twelve days, one of the eight "bionauts," Jane Poynter had to be evacuated for 6.5 hours to receive medical treatment when a fingertip was accidentally cut off in one of the threshing machines. Even in the week-long simulation before "Closure," the CO2 low was 554 ppm, eerily simulating global warming phenomena occurring in Biosphere I. But John Allen downplayed this problem, with an adamant insistence on the project's success, amounting to a "reign of terror" that also included rejecting the research projecting that only 80% of the food needed by the bionauts could be grown inside the Biosphere (Poynter 2006, 115–116). After 16 months, with carbon dioxide levels rising up to a high of 4,500 ppm, seven tons of oxygen "missing," and oxygen levels falling under 15%, causing the resident medical doctor becoming unable to add up simple columns of numbers, an emergency situation was declared and additional oxygen was pumped back in (Cooper 1991d). Despite working sixty-six-hour weeks to produce food and maintain

Biospheric operations, the eight Biosphereans were able to produce only 80% of the food needed for their subsistence, as predicted (and suppressed) prior to closure, and although their nutrient levels remained sufficient, their bodies lost weight, sleep, and strength. The Biosphereans fell into two warring groups before the first year was out: one group insisted on reporting and responding to the real scientific data, while the other group remained loyal to John Allen and his vision, regardless of the material, biological data from scientific instruments and their own animal bodies. Tensions between the two factions ran so high that from month ten through the remaining two-year enclosure, Biosphereans passed one another in the narrow hallways by averting eyes and hugging the wall (Cooper 1991d). Tensions persisted to such a degree that when a second "mission" for a six-month enclosure with seven "bionauts" was launched on March 6, 1994, two members of the first mission traveled from Japan to Oracle, Arizona to break the seals of Biosphere II at 3:00 a.m. on April 5. Their break-in occurred three days after Ed Bass seized control of the project from John Allen and Margret Augustine, who had been running Biosphere II by mismanaging finances by millions of dollars, and rejecting scientific advice. As of June 27, 2011, the University of Arizona has taken over management of Biosphere II, now termed "B2, Where Science Lives."

What evidence suggests Biosphere II was antiecological? First, the purpose of the mission was colonizing and capitalizing on outer space, not solving environmental problems here on earth. In their introductory chapter of *Space Biospheres*, Allen and Nelson explain that their purpose is "to assist the Biosphere [earth's ecosystems] to evolve off planet Earth into potential life regions of our solar system" and respond to the "historic imperative" of colonizing Mars, given the "inevitable doom" of Earth (Allen and Nelson 1989, 3). An odd part of that imperative, months prior to closure of Biosphere II, appeared on May 15, 1991, when plans for commercial development of the 3,600 acres around Biosphere II were submitted to local planning officials and included opening up RV parks, shopping centers, gas stations, offices, schools, hotels, apartments, and a golf course (Cooper 1991c). The proposed community would include areas for research and development (the Biosphere II building) and ecological public education that would include environmental interpretive centers, learning institutions, technical schools, and accommodations for students, scholars, individuals, and families. The unmasking of techno-science as capitalist commercial venture is seldom so clear.

Second, the mechanistic approach to recreating Biosphere 1—tokenism guided by anthropocentrism, selecting the nearly 4,000 species for inclusion based primarily on their functions that benefit humans—is fundamentally antiecological and unsustainable (as outcomes from the two enclosures demonstrated). Our planet's ecosystems and inhabitants interact in ways and on scales still not fully understood by human scientists or material philosophers alike, whose theories about "vibrant matter" and the earth's "dense network" of agencies (Bennett 2010) have yet to distinguish *right relations* (i.e., ecologically sustainable and socially just) among those agencies, and who tend to ignore interspecies relations (especially between humans and other animal species) altogether. Attending to the sustainability of these diverse ecological "intra-actions" is crucial, for as even its critics agreed, Biosphere II's most important lesson is that there is no alternative to Earth.

Another lesson involved food: despite attempts in advance planning, the food systems in Biosphere II assumed the deaths of nonhuman animal species were a requisite part of the human diet. One source reported being surprised at the "belated realization that we had to farm organically" because such a closed system would be "required for permanent bases in far-away places such as Mars"; the agriculture was going to eschew "green revolution" technologies and go organic, in recognition of how foundational organic agriculture is to ecological sustainability (Poynter 2006, 95, 182). But human-nonhuman animal relations were given no consideration in the Biospherean diet, and were not seen as relevant to a new ecological vision; hence, the repeated slaughter and consumption of animals whom the Biosphereans had regarded as friends clearly diminished their ecological ethics. Poynter recalls, "When in the animal bay, I often thought about how we received all this wonderful milk, eggs, and meat essentially for free. The miniature chickens, pigs, and goats lived off stuff we could not eat" (183). It never occurs to her (or the other Synergists) to consider the cost of the animal's life to that animal was far from "free," nor did these animals choose to enter the Synergists' experiment (theater) of Biosphere II. As the Biospherean in charge of animal agriculture, Poynter found it "harder and harder to butcher the animals" as she began "living on a mostly vegetarian diet," because she "felt even more connected to the [animals] once [she] knew [she] would not be eating them" (184). Once the food shortages became evident, the Biosphereans decided to eat the pigs that had been their companions, as Poynter reports: "I was sad to see Zazu and Quincy go. It felt like a betrayal to eat them. They had been with us for several years, and it was like eating a friend" (228). These insights were short-lived.

A fourth antiecological feature was that human social culture was given insufficient consideration: the fact that all the Biosphereans were white, heterosexual (or celibate), and largely from privileged backgrounds was not seen as a concern, nor was the concern that future biospheres would also be available only to a small group of (presumably elite) humans, as *Space Biospheres* explains: the "first Mars Base . . . will be corporate in form . . . the population can range from 64 to 80 people. If more population arrives they will have to begin their own communities" (Allen and Nelson 1989, 7). Where these newcomers will find another billionaire to finance their personal Biosphere is not stated. Interpersonal relationships were expected to be subsumed to the group, placing holism over individuals, an ethical strategy strongly criticized by feminists for the ways that it devalues loving partners, children, families, friendships, and individuals as well. Children were not built into the plans for time, energy, or nurturance in the Synergia community that preceded Biosphere II, and couples in both communities were expected to maintain their relationships outside of the times allotted for community work and activities (Poynter 2006; Veysey 1973).

Given all these flaws, why did Biosphere II succeed to attract attention and credibility as long as it did? Certainly the millions of dollars in funding from Ed Bass, along with the purchase of scientific individuals and organizations gave the project visibility and credibility.[14] The people behind the project took cover behind a diversity of international, incorporated entities—Synergia Ranch (New Mexico), Institute of Ecotechnics (London), Caravan of Dreams Theater (Texas), Space Biosphere Ventures (Arizona), Decisions Investment Team—all staffed by the same people and controlled by the same core group. The group also managed the public media very effectively, and hired lawyers to use threats of litigation when the media coverage was unfavorable. But all these strategies would not have succeed outside of the encompassing cultural ideologies of masculinist techno-science (notably its corollary beliefs that science and technology will save humanity from any crisis, even providing alternatives to this world if we end up trashing the planet) and neoliberal economics (encapsulated in the slogan that "What's good for GM is good for the country," that is, what is good for an elite few/corporation is good for the nation, and the earth as well; moreover, if a project/person/organization has a lot of money, he/it must have "done something right" and thus be credible and trustworthy). In sum, Biosphere II's experiment confined animals of diverse species (including humans) in a two-year "spaceship" demonstrating that our animal "entanglement" with the earth's ecosystems cannot be mimicked without severe damages to animal and ecosystem health. Confronting the causes of global climate change, we need to learn from and reject these antiecological beliefs.

MASCULINIST CULTS, SPACE ESCAPES, AND OTHER TECHNO-SOLUTIONS FOR CLIMATE CHANGE

Beginning with the space race of the 1950s and 1960s, and fueled by the experiments with Biosphere II in the 1980s and 1990s, space exploration has been powered by two forces: a white, masculinist conception of techno-science, and huge amounts of money—sourced from both government funding and private investments. In 1970, Gil Scott-Heron's lyrics to the popular "Whitey on the Moon" pinpointed the diversions, asking "Was all that money I made las' year/(for Whitey on the moon?)/How come there ain't no money here?" as shifting federal tax funds from social well-being (public education, health and welfare) to outer space projects effectively shifts benefits from working and middle class taxpayers to economic elites.

As early as 1967, Barron Hilton, president of Hilton Hotels, envisioned putting hotels in space, and similar proposals in the 1970s from Princeton physicist Gerry O'Neill for human habitations in space are now cited by NewSpace proponents as prescient inspirations (Dickens 2009; Valentine 2012). Coined by the Space Frontier Foundation (SFF) in 2006, the term "NewSpace" primarily refers to wealthy entrepreneurs who have launched corporations with names like SpaceX, Virgin Galactic, XCOR, and Bigelow Aerospace, with the primary purpose of designing and promoting space tourism independent of NASA. Their success seems immanent: in late 2010, Virgin Galactic conducted its first landing of WhiteKnightTwo at Spaceport America in New Mexico, with plans to fly customers to suborbital space by 2013; SpaceX's Falcon 9 rocket with its Dragon Space Capsule was launched in 2010 and on May 25, 2012, the Dragon successfully docked at the International Space Station (Valentine 2012).

What motivates these NewSpace advocates? According to Peter Dickens, the cosmos has become capitalism's new "outside," and these "outer space imperialisms" are now seeking "outer spatial fixes"—investments in outer space—to solve the crises of capitalism (Dickens 2009, 68). President Eisenhower's neologism of the "military-industrial complex" has become the "military-industrial-*space* complex" inventing new enemies that require increased surveillance and funding for defense contractors such as Raytheon, General Dynamics, Lockheed Martin, Boeing, and Northrop Grumman. The interimperialist rivalries from the Cold War have expanded the world into three power blocs competing for outer space: the USA (whose Department of Defense philosophy is called "Full Spectrum Dominance"), Europe, and China. According to Dickens, three arguments are used to legitimate "outer spatial fixes": appeals to the "pure, universal, scientific knowledge to be supposedly gained" by outer space exploration; benefits to the global environment and world population, including "monitoring" of ecological conditions, collecting solar energy for a world running out of resources, and "protecting" citizens' freedom; and fulfilling the biologically-engrained need of humanity to "explore," and "conquer new horizons," releasing the "human potential" that enabled earlier colonialist ventures (Dickens 2009, 78–79). The gendered and colonialist rhetoric of these arguments needs little commentary: they present science as value-free and acontextual, and scientific knowers' identities are constructed via rugged individualism and conquest, all features of masculinism; and these arguments appeal to fear, satisfying a false need for more "monitoring" when global monitoring already confirms the ecological conditions of a climate change crisis (i.e., melting polar ice, increasingly severe weather events, record-breaking heat, drought, species migrations, and extinctions).

Moreover, using outer space to collect solar energy for a world running out of resources presumes we have exhausted our capacities to collect solar energy here on earth, when this is far from accurate; however, this assumption does express the ideology of NewSpace in its rejection of "limits to growth" positions popular since the 1970s. According to NewSpace advocates, space has boundless amounts of energy, fuel, minerals, and land mass; it can provide space-based solar power, metals from mining asteroids, and expanded free markets (Valentine 2012). Attending the conferences of NewSpace advocates, David Valentine found three subgroups, each with a different perspective on the purposes of space exploration. At the Space Investment Summits conference, Valentine heard frequent iterations of "space is expensive," from investors primarily interested in the "exit strategy," or point at which a business can be sold and investors can reap their profits. But at the National Space Society's International Space Development Conference (ISDC), the slogan was "Space is a place, not a program": here, advocates see space as a "privileged destination" because "the species depends on it" (Valentine 1050–57). This view leads to Valentine's third group, which measures the success of NewSpace by the point at which "humans don't have to return to Earth."[15] These suggestions are eerily resonant with images of Laika in Sputnik 2's no-return voyage of 1957, or Stanley Kubrick's *2001: A Space Odyssey* (1968) concluding image of a human fetus floating in outer space, without mother or womb or earth for food, warmth, and nurturance. Such images of the future are not "astroenvironmentalisms" (Henry and Taylor 2009)[16] but Icarian, hubristic antienvironmentalisms infatuated with the sublime, defined by Edmund Burke as vastness, darkness, infinity, vacuity, difficulty and danger, confronting us with our mortality and our insignificance in relation to something much greater than ourselves (Smith 2009). As Patrick D. Murphy (2012) has ably argued, the sublime is antithetical to an ecofeminist environmental ethic.

Here on earth, techno-scientific attempts to mitigate the pace and effects of climate change are being undertaken by heroic entrepreneurs operating outside the bounds of government. Geoengineering is now attempting to substitute for the real and difficult work of reducing emissions; bringing corporations and governments in line with real climate science facts, and creating policies affecting the behaviors and economics at all levels (governments, corporations, communities, individuals); and ultimately adapting to and seeking to mitigate the unavoidable effects of climate change already occurring. After the NASA space race and Biosphere II, huge amounts of money are still deciding national and international responses to climate change. Who benefits from such denials of ecological science and the ecological humanities, and who pays for those benefits?

As Naomi Klein (2012) explains, geoengineering involves "high-risk, large-scale technical interventions that would fundamentally change the oceans and skies in order to reduce the effects of global warming." The strategies being considered include "pumping sulfate aerosols into the upper atmosphere to imitate the cooling effects of a major volcanic eruption and 'brightening' clouds so they reflect more of the sun's rays back to space." Today, backed by the U.S. House Committee on Science and Technology, the British Government, and billionaire Bill Gates, scientists are preparing to "actively tamper with the complex and unpredictable natural systems that sustain life on earth—with huge potential for unintended consequences" (Klein 2012). The most frightening features of geoengineering are that earth's systems are connected in ways scientists still do not fully understand (witness Biosphere II) so that geoengineering efforts in one part of the globe could trigger disastrous outcomes in another part of the globe—and there's no oversight mechanisms in place. Unlike the UN Framework Convention on Climate Change (UNFCCC), which proposes a community-wide, nation-by-nation commitment to lower GHG emissions, any individual or group with the will and the funding can attempt a geoengineering "solution."

Like shooting chimps into space, confining Biosphereans and their "food animals" inside a glass dome, or creating hotels and shuttles for tourists in NewSpace, geoengineering follows the same misguided assumptions that have brought us to the current climate crisis: the belief that humans are somehow separate from and above nature, and humans must control nature. This humanist cultural belief is a deeply Euro-Western articulation of heteromasculinity whose key characteristic is dominance—physical, economic, political, military, ecological, psychological, emotional, and sexual dominance. Feminists and anthropologists have described these colonialist Euro-Western cultures as "warrior *cults*" shaped and inflected by the assumptions and ideology of cultural heteromasculinity. If "we" are not above and in control of nature—whether via imperialism of other nondominant people, places, and species, or via techno-scientific animal experimentation under the guise of space exploration—then "we" cease to be "real men" and thus cease to be human, becoming not just "humananimals" (Haraway's term) but more specifically, *earthanimals*. As this chapter demonstrates, animals in space are, ultimately, dead animals.

For millennia, humans of all civilizations and cultures have looked to the skies with awe, interest, and curiosity. Some have crafted religions, others have sought to read human destinies, and still others simply measured the movements of time or their own oceanic voyages by watching the stars. Somehow, through a nexus of social, economic, and ecological forces, some humans lost understanding of ourselves as but one animal species among many, of our shared place in the universe and on this one Earth. Rather than face our *entanglement* with the rest of nature (Alaimo 2010), and the strengths and limitations of our own *earthanimalities*, outer space advocates pursue techno-scientific solutions in the Anthropocene when our future depends on confronting and reducing the causes of climate change itself: industrial, agricultural, and transportation processes and productions, including deforestation and animal-based food production, that are increasing GHGs via first world overconsumption habits, as climate justice activists from Doha to Detroit agree.

CONCLUSION: TOWARD ECOMASCULINITIES ON EARTH

In *Nature Ethics*, Marti Kheel argues that the social construction of dominant masculinity is inherently antiecological for the ways it "idealizes transcending the [female-imaged] biological realm, as represented by other-than-human animals and affiliative ties" and "subordinate[s] empathy and care for individual beings to a larger cognitive perspective or 'whole'" (2008, 3). Of major significance is Kheel's insight that *all environmental ethics are constructed through the lens of gender.* If environmental ethicists and activists want to make more conscious choices about that lens, particularly in the ways that it influences the environmental sciences and humanities, economics and politics, then we'll need to envision more sustainable, just, and diverse expressions of ecogenders, ecomasculinities, and ecosexualities, as I explain in chapter 8. Already, the climate justice movement has benefited from the new social movements and radical environmentalisms of the late twentieth century, but as even the internationally acclaimed 350.Org shows, there's still room to grow.

On November 30, 2012, Bill McKibben's "Do the Math" tour made a stop in Minneapolis to update the local branch of climate justice activists, MN350.Org, on the challenges and next steps for the climate justice movement. Introduced by folk singer Mason Jennings, with presentations from Marty Cobenais of the Indigenous Environmental Network (IEN), Polar explorer Will Steger, and Winona LaDuke of the White Earth Land Recovery Project (WELRP), Bill McKibben's talk featured Minnesota-based video of our grassroots activisms, complemented with video of the global 350.Org movement, beginning in 2007 with "Step It UP" through the Copenhagen

Convention and beyond. To save transportation costs, McKibben interspersed his talk with taped interviews from Van Jones of Green For All, an organization to develop a green economy that lifts people out of poverty; the producer of "Gasland" documentary, Josh Fox, describing the human and ecological effects of fracking, and the inadequacy of "backyard" or local ecoactivisms without an end to climate change; and Archbishop Desmond Tutu from South Africa speaking about apartheid and the divestment strategies of the 1980s. Contrasting the global warming evidence provided by NASA scientist James Hanson, versus the global warming deniers' pseudoscience funded by oil companies and their think tanks, McKibben demonstrates a more feminist *ecomasculine* approach to scientific knowledge-construction in his methods of building a grassroots and global environmental movement with racially and nationally diverse leaders, his methodologies of encouraging a strong sense of participatory democracy, and his passionate epistemology, which involves listening to and creating community conversations among all those involved in a climate justice movement that benefits all participants (McKibben 2013). McKibben's "math" equation is simple: $CO2 + \$ =$ a burning planet (McKibben 2012). Accordingly, the next step is subtraction: 350.Org plans to encourage international strategies of divestment, withdrawing college and university investments from global oil corporations like ExxonMobil, Shell, ChevronTexaco, BP, and ConocoPhillips.

McKibben's inclusive and democratic approach to building "Movements Without Leaders" (2013) could be even stronger by attending to the intra-action between species justice and climate justice.[17] Though McKibben mentioned "and other species" several times in his talk, his agnostic position on human-other animal relations (McKibben 2010) where he admits it's "pretty clear" that eating less meat is a good idea, but "we don't really take official stances on issues like veganism." Ignoring the links between animal-based agriculture and climate change seems ludicrous to James McWilliams, author of *Just Food: Where Locavores Get It Wrong and How We Can Eat Responsibly* (2010). As McWilliams speculates, there are at least three reasons for the omission, and they aren't flattering: first, getting arrested in front of the White House for opposing the Tar Sands Pipeline models an ecoheroic (and masculinist) stance that garners headline coverage and is "a lot better for 350.org's profile than staying at home, munching kale, and advising others to explore veganism" (McWilliams 2011). Moreover, pipelines provide the media with clear victims, perpetrators, and a narrative of ecological decline that is less visible than the ongoing first world overconsumption of intensely farmed animals and their associated ecological impacts—another example of "slow violence" that is harder to make visible (Nixon 2011). Second, meat-eating environmentalists who argue that we must replace feedlot farming with rotational grazing, as McKibben does, nostalgically refer to a preindustrial and preagrarian past, implying that nature is more natural in the absence of human beings. In doing so, they reiterate an entrenched human/ nature dualism that persists among diverse branches of environmentalisms, despite incisive critiques from posthumanist, ecofeminist, material feminist, and other philosophies. Finally, as McWilliams argues, meat eating seems to represent "personal freedom" and individual choice, while oil pipelines and coal power plants offer more visible and collectively shared images of environmental impact, a contrast that articulates differences between rights-based ethics and the more feminist relational ethics of care and responsibility. But the effects of animal-based food consumption as well as unsustainable energy and transportation are all contributing to climate change, and some scientists suggest that a change in diet may be as crucial as stopping an oil pipeline: study after study shows that steep reductions in livestock production, along with returning the world's pastures (a quarter of the land surface) to grow trees, woodland and native perennial grasses, can soak up carbon emissions and combat climate change (FAO 2006; Harvey 2016; Springman et al. 2016). Why would 350.Org overlook this crucial component of climate justice ecoactivism? Why would anyone not profiting from industrial agribusiness overlook the influence of gender and species on environments and climate alike?

In the necessary move to replace anti-ecological masculinist approaches to the environmental sciences and humanities with more ecological masculinities, we can find seeds of this transition in even the masculinist hunter-environmentalists Kheel has criticized. As Aldo Leopold wrote in *A Sand County Almanac*, "a land ethic changes the role of *Homo sapiens* from conqueror of the land-community to plain member and citizen of it," a change Leopold urged because "the conqueror role is eventually self-defeating."[18] In the role of conqueror, science claims to know "what makes the community clock tick," but in fact "the biotic mechanism is so complex that its workings may never be fully understood." Leopold's land ethic defines a set of paradoxes, and offers humans the linked choices that involve Western cultures' humanist identity, our use of science, our environmental ethics, and our society's rejection of racism and classism: will our culture continue to be "man the conqueror," or can we become "the biotic citizen"? Will science be "the sharpener of [the conqueror's] sword" or "the searchlight on [the] universe"? And will the earth itself, its interdependent ecosystems, plants, animals, and human communities, become the conquerors' "slave and servant" or "a community to which we belong"?

If we "do the math," the relevance of these questions to climate science, climate justice, interspecies relations—and our own ecopsychology—become evident.

NOTES

1. It's more than coincidence that astronauts Neil Armstrong and James Lovell have both spoken of blotting out the earth with their thumbs. The first man on the moon, Armstrong reportedly said, "It suddenly struck me that that tiny pea, pretty and blue, was the Earth. I put up my thumb and shut one eye, and my thumb blotted out the planet Earth. I didn't feel like a giant. I felt very, very small" (see Neil Armstrong's entry in the New Mexico Museum of Space History's International Space Hall of Fame http://www.nmspacemuseum.org/halloffame/detail.php?id=1). Similarly, the astronaut famous for his excursions on Apollo 8 and 13, Lovell commented on "the fact that just from the distance of the Moon you can put your thumb up and you can hide the Earth behind your thumb. Everything that you've ever known, your loved ones, your business, the problems of the Earth itself—all behind your thumb" (see quotes from the documentary "In the Shadow of the Moon" at http://www.imdb.com/title/tt0925248/quotes).

2. Haraway's understanding of species dominance is limited by her own human-ism, and this limits her theorizing considerably (see Weisberg 2009).

3. As Kheel noted, "researchers do not 'kill' animals in laboratories; the word 'sacrifice' is still employed" (1989, 104). Kheel critiqued patriarchal religious myths of a Father-God who offers his son to die in exchange for others' redemption, or who asks His followers to kill their most precious son as proof of their devotion—but who will be appeased by the killing of another animal's offspring instead. For an overview of the linguistic battles between vivisectors and animal advocates, see Gruen 2011.

4. One notable exception was provided by U.S. Air Force Surgeon Major John Paul Stapp, who created the "Gee-Whizz" sled in 1947, and following test runs with 185-pound mannequins (later dubbed "crash test dummies" by the auto industry, which used these in safety tests after being compelled to do so by the federal govern-ment), rode the sled himself with acceleration forces of 45 g's and survived without lasting injuries; he went on to test the sled on chimpanzees a total of 88 times, with some at a "crushing 270 g's" of force, leaving the animal's body "a mess" (See Bur-gess and Dubbs 2007, 103).

5. The actual lines from *Hamlet* read thus: "Alas, poor Yorick! I knew him, Horatio; a fellow of infinite jest, of most excellent fancy; he hath borne me on his back a thousand times; and now, how abhorred in my imagination it is! My gorge rises at it. Here hung those lips that I have kissed I know not how oft. Where be your gibes now? Your gambols? Your songs? Your flashes of merriment, that were wont to set the table on a roar?" (*Hamlet*, V.i) The correlation of a court jester and a space chimp most likely sent to his death underscores the "scientific" mockery that was made of these animal lives.

6. Forty years later, a website devoted to Moscow's homeless animals still tells the story of Laika; see http://www.moscowanimals.org/index.html accessed on 11/26/2012.

7. After the Air Force denied Noon's requests, awarding even more space chimps to the Coulston Foundation, Noon sued the Air Force for custody and raised funds to build Save the Chimps Sanctuary in Florida, where these chimps can enjoy a life

free of testing in an outdoor refuge with islands, play areas, and fresh healthy foods. (See http://www.spacechimps.com/theirstory.html) But animal aerospace testing did not end. The American Anti-Vivisection Society reports that through 1996, NASA was still conducting a multimillion dollar research project called Bion that involved sending monkeys whose tails were cut off and who were placed into apparel similar to straight-jackets with restraining rings screwed into their skulls and various electrodes implanted throughout their bodies into space for 14 days. The purpose of Bion was to study the effects of microgravity and radiation in living beings. These flights ended through a confluence of forces involving the deaths of some space monkeys and the persistent efforts of animal rights groups pressuring Congress and NASA.

8. At the same time, heroic and comic narratives continue to be produced as attempts to obscure these historical facts: that is, Richard Hillard's *Ham the Astro-chimp* (2007) and the Disney movie, "Space Chimps" (2008).

9. Backgrounding is one of the five operations of dominance constructing Val Plumwood's (1993) explication of the Master identity; the other operations include hyperseparation, incorporation, instrumentalism, and stereotyping.

10. In his excellent study of fatherhood across species, Jeffrey Moussaieff Masson (1999) argues persuasively for nurturance as a crucial characteristic for human fathers, observing the many varieties of fatherhood behaviors across species, and the ways that the behaviors and norms for patriarchal fatherhood (exaggerated in the scientists' treatment of animals used in space exploration) are culturally distorted and enforced by social institutions.

11. For a critique of this delayed uptake, see Gaard 2012.

12. Though NASA's budget has remained at or below 1.0% of the U.S. federal budget since 1958, with the exception of the moon-race era, 1962–1972, that 1.0% amounts to billions of dollars. See http://www.guardian.co.uk/news/datablog/2010/feb/01/nasa-budgets-us-spending-space-travel accessed 11/20/2012.

13. Margret Augustine was pregnant with John Allen's child during the nine months prior to Biosphere II Closure. The charisma of Allen extended to his love life, of course, involving his wife Marie Harding, who cashed in her entire inheritance to provide the down payment for Synergia Ranch, and later Cathleen Burke, Allen's lover for the ten years she spent in his group, and who reported "beatings" in which Allen "beat mostly the core members of the group" as well as Burke and even his funder, Ed Bass (Siano 1992).

14. Research funding was given to Dr. Ghillean Prance, director of the Royal Botanical Gardens at Kew, England, for setting up the rain forest in Biosphere II; the Yale School of Forestry and Ecological Science was given $20 million to create the Yale Institute of Biospheric Studies, and $40,000 to work with Biosphere II on "carbon budgeting"; the National Center for Atmospheric Research, funded through U.S. taxpayers via the National Science Foundation, directed $100,000 into Biosphere II (matched by the Biosphere funders); the Smithsonian Institution received at least $400,000 for the consultancy of Dr. Walter Adey and Dr. Thomas Lovejoy; and the Environmental Research Laboratory (ERL) at the University of Arizona was paid $5 million to participate in research leading up to Biosphere II (Cooper 1991a).

15. Valentine's article advocating that the NewSpace adherents be taken seriously—"How do we take this cosmology seriously without thinking that we already know the answer?"—was funded, in part, by the National Science Foundation; see Valentine, 1064–1065.

16. This term is used in Henry and Taylor, 2009, 200. Their idea that we must extend environmental ethics to include and address "space junk" and other polluting particles fits well with a feminist eco-ethic; my concern here is that until we enact genuine environmental justice here on earth, we cannot pretend to be achieving such environmentalisms in space, or to propose those as a replacement or negation of the need for such actions on earth.

17. Emailing McKibben directly to applaud his feminist strategies for movement-building, I was delighted to receive an immediate response from him, directing me to his article on TomDispatch.com where he discusses his intentions in building a "leader-full movement," sustainable like solar power, through "distributed generation" (http://www.tomdispatch.com/blog/175737/). His passionate and inclusive activism for climate justice have built the global movements for climate justice we see in 2016. Including all species will only strengthen this work.

18. Marti Kheel provides a strong and well-supported critique of Leopold's masculinism (which involved lifelong hunting) in her *Nature Ethics: An Ecofeminist Perspective* (2008). At the end of his life, Leopold continued to evolve his environmental ethics, and in "The Land Ethic" his writing contains implicit acknowledgements of the links among diverse kinds of oppression—gender, sexuality, race, class, and nation. Leopold was radically ahead of his time in challenging the very nature of human identity as linked to environmental behaviors and relationships, although such ideas and language were not available to him in 1948.

Part III

CLIMATES

Chapter 6

Climate Justice

Since the times of Ancient Rome, Lady Justice has been depicted wearing a blindfold representing objectivity, holding scales to weigh competing claims in her right hand, and a sword of reason in her left hand. Contemporary feminist justice ethicists have critiqued the masculinist bias of traditional Western ethics for the ways it overvalues reason and objectivity, devaluing women's standpoints and women's work and envisions justice-as-distribution of resources among discrete individuals with rights, rather than emerging through relationships which shape participant identities and responsibilities (Jaggar 1994; Warren 1990; Young 1990). Ecological feminist ethics have addressed human relationships with other animals, with environments, and with diverse others locally and globally as relations meriting contextualized ethical concern (Donovan and Adams 2007). But a feminist ethical approach to climate justice—challenging the distributive model that has ignored relations of gender, sexuality, species, and environments—has yet to be fully developed.

To date, climate change discourse has not accurately presented the gendered character of first world planetary overconsumption. For example, a prominent symbol from the Copenhagen Climate Conference of the Parties (COP 15) in December 2009 depicts an obese "Justitia, Western Goddess of Justice" riding on the back of an emaciated black man; in other artworks for the conference, a group of starving African male bodies was installed in a wide river (see Figure 6.1). The image of Justitia was captioned, "I'm sitting on the back of a man—he is sinking under the burden—I will do everything to help him—except to step down from his back" (Sandberg & Sandberg 2010, 8). Allegedly an artwork referencing the heavy climate change burden carried by the global South, and the climate debt owed by the overconsuming global North, from a feminist perspective the missing critique is that the genders are reversed: women produce the majority of the world's food, yet the majority of

Figure 6.1   Justitia.

the world's hungry are women and children, not men. And the overconsumption of earth's other inhabitants—plants, animals, ecosystems—is not even visibly depicted.

In this chapter, I argue that climate change and first world overconsumption are produced by masculinist ideology, and will not be solved by masculinist techno-science approaches. Instead, I propose, queer feminist posthumanist climate justice perspectives at the local, national, and global levels are needed to intervene and transform both our analyses and our solutions to climate change.

HERSTORY: WOMEN'S CLIMATE CHANGE ACTIVISM

Although the "first stirrings" of women's environmental defense were introduced at the UN 1985 conference in Nairobi, through news of India's Chipko movement involving peasant women's defense of trees (their livelihood), women's role in planetary protection became clearly articulated in November 1991, when the Women's Environment and Development Organization (WEDO) organized the World Women's Congress for a Healthy Planet in Miami, Florida (Resurrección 2013; WEDO 2012). Seen as an opportunity to build on the gains of the UN Decade for Women and to prepare a Women's Action Agenda for the 1992 UN Conference on Environment and Development (UNCED) in Rio de Janeiro, the World Women's Congress drew more than 1500 women from 83 countries. But while its leaders alleged that the resulting "Women's Agenda 21" had been built through a consensus process, for many of those sitting in attendance, listening to one elite speaker after another, it was not clear how our views shaped or even contributed to this process of agenda-formation. Participatory democracy—long a valued strategy in grassroots ecofeminist tactics—was reduced to two dubious threads: a series of break-out discussion groups held throughout the conference, and a "Report Card" for participants to take home and use to evaluate specific issues within their communities and mobilize a local response (shaping the issues themselves had no place on the report card). Along with other ecofeminists, I felt a mix of energy, dismay, and frustration at this gathering.[1] While the women leaders from many countries were valuable participants and decision-makers in the upcoming conversations at the UNCED, that weekend in Miami, too many speakers discussed women's "feminine" gender roles, our "influence" on decision-makers, and the need for "reforms" to the present system—all introduced and capped with the essentializing motto, "It's Time For Women to Mother Earth."

Despite these flaws in rhetoric and democratic participation, WEDO's 1991 World Women's Congress has been hailed as the entry-point for feminism into the UN conferences on the global environment, opening the way for later developments bridging feminist interventions and activisms addressing climate change. The following year, UNCED's Agenda 21 did not in fact include the most transformative recommendations from the Women's Agenda 21—the analysis of environmental degradation as rooted in military/industrial/capitalist economics, for example—or even the more reformist proposals such as implementing gender equity on all UN panels, an issue which has been taken up again at the 2013 COP for the UNFCCC in Warsaw, Poland (See Table 6.1).

Perhaps WEDO's Women's Agenda 21 had already been undermined by the 1987 report from the World Commission on Environment and Development, *Our Common Future*, led by Gro Harlem Brundtland. This report established "sustainable development" as a desirable strategy, defined as "development that meets the needs of the present without compromising the ability of future generations to meet their own needs"—which sounds reasonable enough,

Table 6.1   Comparing Women's Agenda 21 (1991) and the UNCED Agenda 21 (1992)

Issues/Statements	Women's Agenda 21	UNCED Agenda 21
Consumption	• A power that women have to drive industrial development that respects the environment and society • A power that may enable a world alliance to boycott current unsustainable production & consumption models	• Women's role as consumers & impact of their purchasing power affects economies • Implement policies to change unsustainable consumption patterns • New technology can play a role in this process
Technology	• Involves destruction of nature • Has not been within reach of the needs of the poor, nor accessible to women, many of whom have been its victims • Ethical implications of technology & need to democratize it to make it available to & beneficial for women & marginalized groups	• Technology is a benefit in carrying out sustainable development • Reinforce research & promote new technology, and involve developing countries in technological development through knowledge transfer
External debt	• Industrialized countries must admit their exploitation of developing countries' resources • Condemns the negative impact of the IMF and World Bank's restructuring policies, especially on women & children • Proposes paying off external debt, and boycotting banks that uphold it	• Developing countries should pay off their external debt • Incentives for international cooperation to reduce debt were identified
Population	• Main causes of environmental degradation are military & industrial pollutants and capitalist economic systems, not women's fertility rates • Consumption-to-waste ratio per person, which is much higher in the industrialized countries than in poor ones, must be corrected	• Population growth is an unsustainable environmental pressure • Family planning policies and educational programs for women are needed • Raise the educational level of women • Promote women's economic independence & participation at decision-making • War on poverty is a key factor in reducing demographic growth

(Data *Source*: Brú Bistuer & Cabo 2004.)

until one reads the document's renewed call for continued economic growth on a finite planet, a fundamentally unsustainable endeavor. The report completely omits discussion of the First World/North's[2] over-development and its high levels of production, consumption, and disregard for the environment (Agostino and Lizarde 2012). Nonetheless, the Brundtland Report's "sustainable development" concept has shaped climate change discourse for the subsequent decades, producing techno-solutions such as "the green economy" that have perpetuated capitalist and colonialist strategies of privatization, and fail to address root causes of the climate crisis (Pskowski 2013).

In the two decades since WEDO's Women's Agenda 21, feminist involvement in global environmentalism has developed from a 1980–1990s focus on "women, environment and development" (WED), "women in development" (WID) or "gender, environment and development" (GED) to an emphasis on feminist political ecology in the 1990s–2000s (Goebel 2004; MacGregor 2010; Resurrección 2013). Initially, discussion of women and environment focused on women in the global South, whose real material needs for food security and productive agricultural land, forest resources, clean water, and sanitation trumped more structural discussions about gendered environmental discourses (i.e., Leonard 1989; Sontheimer 1991), although these structurally transformative elements were equally present in other texts (i.e., Sen and Grown 1987). The focus on *women* rather than *gender* tended to construct women as *victims* of environmental degradation in need of rescue; their essential closeness to nature, cultivated through family caregiving and through subsistence labor, was argued as providing women with special knowledge, and their *agency* as laborers and leaders in environmental sustainability projects was advocated (Shiva 1989; Mies and Shiva 1993). Clearly, this rhetoric instrumentalized women and ignored the cultural limitations of the woman-nature linkage (cf. Li 1993; Dodd 1997; Leach 2007); it was also significantly silent on the roles of men, and the ways that gender as a system constructed economic and material resources that produce "victims" (MacGregor 2010; Resurrección 2013). The shift to a "feminist political ecology" (Goebel 2004) involved a macrolevel exploration of the problems of globalization and colonization, a microlevel examination of local institutions for their environmental management, a critique of marriage institutions for the ways these affect women's access to natural resources, and an interrogation of the gendered aspects of space in terms of women's mobility, labor, knowledge, and power. The shift from *women as individuals* to *gender as a system structuring power relations* has been an important development in feminist responses to climate change.

Moving forward from this herstory, I bring an ecofeminist perspective to examine the ways that climate change phenomena have been analyzed primarily from the standpoint of the environmental sciences and technologies, and how this standpoint forecloses the kinds of solutions envisioned. I examine both liberal and cultural ecofeminist perspectives highlighting the ways women have been both excluded from climate change policy discussions and disproportionately affected by climate change phenomena, and summarize proposals drawing on women's "special knowledge" and agency as decision-makers and leaders in solving the problems of climate change. Noting the popular utility as well as the limitations of these perspectives, I examine both climate change phenomena and climate justice analyses. In organizing this inquiry, I am inspired by feminist activist and scholar Charlotte Bunch, founder of Rutgers University's Center for Women's Global Leadership, whose landmark essay, "Not By Degrees: Feminist Theory and Education" (1979) proposes four tactical steps for using feminist theory to understand situations, place them in a broader context, and evaluate possible courses of action. Simply stated, Bunch's theory suggests we ask, What is the problem? How did it originate? What do we want? And, how do we get there? (Bunch 1987).

WHAT'S THE PROBLEM? CLIMATE CHANGE, ENVIRONMENTAL SCIENCE, AND REFORMIST FEMINISMS

The scientific evidence of climate change should be alarming: since the Industrial Revolution (variously dated as beginning between 1760 and 1840), when the density of carbon dioxide in the atmosphere was just 280 parts per million (ppm), humans began burning coal, gas, and oil to produce energy, provide transportation, and fuel machineries. Carbon dioxide increased gradually until 1900, when GHGs and global temperatures began to skyrocket, as shown in Michael Mann's "hockey stick" graph included with the 2001 Intergovernmental Panel on Climate Change (IPCC) summary for policymakers (Appell 2005). Fast forward to the summer of 2012, by which time half of the Arctic Sea ice had vanished. In May 2013, Hawaii's Mauna Loa Observatory recorded carbon dioxide levels at 400 ppm, exceeding all historical records, and continuing to increase at a pace exceeding 2 ppm per year. The ecological consequences of climate change—rising sea levels, melting ice sheets and receding glaciers, vanishing coral reefs, extreme weather events (i.e., hurricanes, floods, droughts, wildfires, heat waves), accelerated species migrations or extinctions, the spread of insect-borne diseases—are already evident. Produced by the planet's most developed countries—with China, the United States, Russia, and India leading the way in highest emissions, and the United States, Australia, Canada, and Saudi Arabia leading with highest per capita emissions—75–80% of the effects of climate change will be felt by the global South/Two-Thirds world, and those effects are most harsh because material poverty means weaker infrastructures of support for housing, clean water, food security, health care, and disaster preparedness/response.

Make no mistake: women are indeed the ones most severely affected by climate change and natural disasters, but their vulnerability is not innate; rather it is a result of inequities produced through gendered social roles, discrimination, and poverty. According to CARE, an international NGO, women work 2/3 of the world's working hours, produce half the world's food, and earn 10% of the world's income; of the world's one billion poorest people, women and girls make up 70%.[3] If there were an unimpeded correlation between hard work and earnings, women would be the world's highest earners. Instead, structural barriers of gender put women—and children—among the world's poorest people, situated on the front lines of climate change. Around the world, gender roles restrict women's mobility, impose tasks associated with food production and caregiving, and simultaneously obstruct women from participating in decision-making about climate change, GHG emissions, and decisions about adaptation and mitigation. In developing countries, women living in poverty bear the burden of climate change consequences, as these create more work to fetch water, or to collect fuel and fodder—duties traditionally assigned to women. When households experience food shortages, which occur regularly and may become more frequent due to climate change, women are the first to go without food so that children and men may eat. As rural areas experience desertification, decreased food production, and other economic and ecological hardships, these factors prompt increased male out-migration to urban centers with the promise of economic gain and wages returned to the family; these promises are not always fulfilled. In the short term, and possibly long term as well, male out-migration means more women are left behind with additional agricultural and household duties, such as caregiving. These women have even fewer resources to cope with seasonal and episodic weather and natural disasters.[4]

Gender inequalities mean that women and children are fourteen times more likely to die in ecological disasters than men (Aguilar 2007; Aguilar, Araujo, & Quesada-Aguilar 2007). For example, in the 1991 cyclone and flood in Bangladesh, 90% of the victims were women. The causes are multiple: warning information was not sent to women, who were largely confined in their homes; women are not trained swimmers; women's caregiving responsibilities meant that women trying to escape the floods were often holding infants and towing elder family members, while husbands escaped alone; moreover, the increased risk of sexual assaults outside the home made women wait longer to leave, hoping that male relatives would return for them. Similarly in the 2004 Tsunami in Aceh, Sumatra, more than 75% of those who died were women. In May 2008, after Cyclone Nargis came ashore in the Ayeyarwady Division of Myanmar, women and girls were 61% of the 130,000 people dead or missing in the aftermath (CARE Canada, 2010).

The deaths of so many mothers leads to increased infant mortality, early marriage of girls, increased neglect of girls' education, sexual assaults, trafficking in women and child prostitution. Even in industrialized countries, more women than men died during the 2003 European heat wave, and during Hurricane Katrina in the United States, African-American women—the poorest population in that part of the country—faced the greatest obstacles to survival (Aguilar et al. 2007). Women who survive climate change disasters are then faced with the likelihood of sexual assault: for example, after Hurricane Katrina, rapes were "reported by dozens of survivors" and mentioned in news stories, but there was no discussion of rape support teams being included with the rescue teams, and no mention of reproductive health services that should have been made available to women who had been raped (Seager 2006). Moreover, the likely assaults on gay, lesbian, bisexual, transgendered queer (GLBTQ) persons went unreported

Climate change homophobia is evident in the media blackout of GLBTQ people in the wake of Hurricane Katrina, an unprecedented storm and infrastructure collapse which occurred just days before the annual queer festival in New Orleans, "Southern Decadence," a celebration that drew 125,000 revelers in 2003 (ecesis.factor). The religious right quickly declared Hurricane Katrina an example of God's wrath against homosexuals, waving signs with "Thank God for Katrina" and publishing detailed connections between the sin of homosexuality and the destruction of New Orleans. It is hard to imagine GLBTQ people not facing harassment, discrimination, and violence during and after the events of Katrina, given the fact that Louisiana, Alabama, and Mississippi lack any legal protections for GLBTQ persons and would have been unsympathetic to such reports.

Queer and transgendered persons already live on the margins of most societies, often denied rights of marriage and family life, denied health care coverage for partners and their children, denied fair housing and employment rights, immigration rights and more. Climate change exacerbates pressures on marginalized people first, with economic and cultural elites best able to mitigate and postpone impacts; as a global phenomenon, homophobia infiltrates climate change discourse, distorting our analysis of climate change causes and climate justice solutions, and placing a wedge between international activists. For example, at the First Worldwide Peoples' Conference on Climate Change and Mother Earth held in Cochabamba, April 19–22, 2010, Bolivian President Evo Morales claimed that the presence of homosexual men around the world was a consequence of eating genetically modified chicken: "The chicken that we eat is chock-full of feminine hormones. So, when men eat these chickens, they deviate from themselves as men" (ILGA

2010). This statement exemplifies a dangerous nexus of ignorance, specie-sism, and homophobia that conceals the workings of industrial agribusiness, and simultaneously vilifies gay and transgendered persons as "genetic devi-ants." Yet in statements of climate justice to date, there is no mention of the integral need for queer climate justice—although all our climates are both gendered and sexualized, simultaneously material, cultural, and ecological.[5]

Described largely from the perspective of the environmental (climate) sciences (i.e., astrophysics, atmospheric chemistry, geography, meteorology, oceanography, paleoclimatology), climate change has been most widely dis-cussed as a scientific problem requiring technological and scientific solutions without substantially transforming ideologies and economies of domination, exploitation, and colonialism: this misrepresentation of climate change root causes is one part of the problem, misdirecting those who ground climate change solutions on incomplete analyses (cf. Klein 2014). On an interna-tional level, solutions mitigating climate change include Reducing Emis-sions from Deforestation and Forest Degradation (REDD+ Initiative), the Kyoto Protocol's Clean Development Mechanism (CDM) that encourages emissions trading, sustainable development funding for Two-Thirds coun-tries, genetically modified crops, renewable energy technologies, and the more recent strategy, geoengineering (Klein 2012). On an individual level, citizen-consumers of the North/One-Thirds world are urged toward green consumerism and carbon-footprint reduction. Certainly, renewable energy is a necessary and wholly possible shift; moreover, it carries within its practice the ideological shift needed to make a wider transformation in the North/One-Thirds consumers' relationship with environments and ecosystems. From a feminist perspective, however, the problem remains that at the highest levels of international discussion, "climate change is cast as a human crisis in which gender has no relevance" (MacGregor 2010) and "man" is supposed to mean "everyone." Such gender-blind analysis leads to excluding data and perspec-tives that are crucial in solving climate change problems, while the issues that women traditionally organize around—environmental health, habitats, liveli-hoods—are marginalized by techno-science solutions which take center stage in climate change discussions and funding. GLBTQ issues such as bullying in the schools, hate crimes legislation, equity in housing and the workplace, same-sex marriage (not to mention polyamorous marriage) don't appear in climate discussions either. Given the gender-blind techno-science perspec-tive dominating climate change discussions, queer feminist entry to these discussions has been stalled, trapped between Scylla and Charybdis: over the past two decades, discussions have alternated between the liberal strategy of mainstreaming women into discussions of risk, vulnerability, and adaptation, as WEDO has done; or, adopting the cultural feminist strategy of calling on women's "unique" capacities of caring for family and for environment, women's "special knowledge" and agency based on their location within gender-role restricted occupations, and lauding women's grassroots leadership. In either strategy, "gender" is restricted to the study of women, and feminist analyses of structural gender inequalities that compare the status of men, women, and GLBTQ others are completely omitted.

To date, the UNFCCC "Gender and Climate Change" website addresses these problems by drawing on both reformist liberal ecofeminisms and cultural (essentialist) ecofeminisms. In its statement on women's vulnerability, inclusion, and agency, the UNFCCC website asserts: "It is increasingly evident that women are at the centre of the climate change challenge. Women are disproportionately affected by climate change impacts, such as droughts, floods and other extreme weather events, but they also have a critical role in combatting climate change." In order to perform that "critical role," however, gender parity in climate change discussions is a minimum requirement: women need to be equal members in policy-setting and decision-making on climate change. And to have authentic, inclusive feminism, gender justice and sexual justice must be partnered with climate justice, for women of all genders and sexualities form the grassroots force within these three movements (cf. Olson 2002).

HOW DID THE PROBLEM ARISE? BLAMING OVERPOPULATION AND BACKGROUNDING GENDER ACROSS SPECIES

Misdirecting analyses of root causes, and thus protecting the status quo, three more prominent antifeminist threads companion and vie for prominence alongside the mainstream scientific response to climate change: the linked rhetorics advocating population control, anti-immigration sentiment, and increased militarism. Ever since Paul Ehrlich's *The Population Bomb* (1968), one thread of first world environmentalism has placed overpopulation (primarily in the third world) at the root of environmental degradation, though some more accurate and defensible manifestations of this discourse link population with first world overconsumption, arguing for twin reductions of both. In practice, mainstream population rhetoric has implicitly targeted third world women with "family planning" packages of contraception, abortion, and sterilization, though more recent manifestations of "population science" have been influenced by feminist arguments for reproductive and sexual health/rights as evinced by discussions at United Nations conferences on population in 1974 (Rumania), 1984 (Mexico), and 1994 (Cairo). Arguing that "women and children in poverty are among the most vulnerable to the impacts of climate change, despite their disproportionately low contribution to the problem" (Engelman 2010), the WorldWatch Institute advocates a population reduction approach to the impacts of climate change on the world's most vulnerable communities, implemented through a three-pronged strategy:

• Eliminating institutional, social, and cultural barriers to women's full legal, civic, and political equality with men;
• Improving schooling for all children and youth, and especially increasing educational attainment among girls and women; and
• Assuring that all women and their partners have access to, and full freedom to use, reproductive health and family planning services so that the highest proportion possible of births results from parents' intentions to raise a child to adulthood (Engelman 2010).

While these three strategies may seem globally relevant, they also seem to target populations in developing countries, as evidenced by the WorldWatch Report's cover photo of two women and three children, captioned "A family on their parched land in Niger." The report offers no interviews with the women targeted for family planning to discover whether this strategy is one they desire or would be able to implement, showing a "Father Knows Best" approach to population and climate science.

Approximately 80% of the world's population (the global South) has generated a mere 20% of global GHG emissions: in other words, the other 20% (the global North) is responsible for 80% of the accumulated GHG emissions in our atmosphere (Egero 2013; Hartmann 2009). Despite the clarity of this logic, population reappeared in publications leading up to the 2009 UN Climate Change COP in Copenhagen, with proponents arguing for family planning among poor communities as a cost-effective method of reducing carbon emissions (Egero 2013). Not to be outdone, the UK Population Matters has launched a "population offset" system similar to carbon offsets purchasable by jet-setting first world consumers (MacGregor 2010). On their website, the organization claims that "Pop*Offsets* is the world's first project that offers to offset carbon dioxide emissions through the most cost-effective and environmentally beneficial means—family planning" (see http://www.popoffsets.com/). None of these strategies suggest reducing the North/First World's alarming overconsumption of the planet's resources, or seriously restricting its 80% contribution of GHGs.

Reducing third world population becomes increasingly important when first world overconsumers realize that the severe climate change outcomes already heading for the world's most marginalized communities will create a refugee crisis and urgent migrations of poor people. Since the growing populations of the Two-Thirds World will be hardest hit by climate change effects and will seek asylum in One-Thirds nations—a migration perceived as a threat to the disproportionate wealth (i.e. "security") of the North—the specter of climate refugees has inspired arguments for increased militarization as a protection against migration (MacGregor 2010; Egero 2013). Noting the ways that women are blamed for climate crises which in fact impact women hardest, both during climate disasters and in the frequency of gender-based violence and material hardships following these disasters, Asian Communities for Reproductive Justice (2009) have urged "looking both ways" to recognize the intersections between climate justice and reproductive justice. For all these reasons, feminists have strongly resisted arguments for population as the root cause of environmental degradations, including climate change (Hartmann 1987; Silliman, Fried, Ross & Gutierrez 2004; Gaard 2010b).

Claims about overpopulation in climate change analyses function as an elitist rhetorical distraction from the more fundamental and intersecting problems of gender, sexuality, and interspecies justice. Although many feminist discussions about these issues have remained limited by the perspective of humanism, other millennial feminists at Conceivable Future are leading the way, arguing that

> The climate crisis is a reproductive crisis. For some of us, the perils of climate change have discouraged us from bearing children, foreclosed the opportunity. For others of us, exposure to the fossil fuel industry has already jeopardized our reproductive health, or the health of our children. For some, parenthood has galvanized us towards greater action. For others, the threats to our reproductive freedom have been radicalizing. The climate impacts we see are unfolding during a time of increasing restrictions on reproductive choice and self-determination.[6]

As feminist science studies scholars affirm, the best analysis of the problem of oppression will be the most inclusive—thus, excluding data is not conducive to good research, good argumentation, or good feminism. On this foundation, it is imperative that feminist approaches to climate justice take a material and posthumanist approach by considering the larger environments in which these ethicopolitical problems of climate change are embedded: our interspecies and ecological transcorporeality, manifested in global economics and our practices of global food production and consumption.

Two branches of feminist inquiry support recuperating these "backgrounded" (in Val Plumwood's terms, an operation of the Master Model that supports domination) elements of climate change. Material feminism (Alaimo and Hekman 2008) advances the concept of *transcorporeality*, the physical fact of our co-constituted embodiment with other flows of life, matter, and energy. This recent articulation of feminist theory rests on four decades of feminist science studies and ecofeminist perspectives on the human-environment connection, developing knowledge in the study of gender, race, class, age, and public health. In the 1970s, feminist health advocates began challenging dominant perspectives in science by noting the research focused on male-only samples, and then generalized the results to women and children. These feminists raised questions about women's and children's health by exploring the influence of environment on human health, and exposing environmental links to breast cancer, asthma, lead poisoning, reproductive disorders, and other types of cancers. National women's groups such as Silent Spring Institute and Breast Cancer Action have worked to bring a feminist environmental perspective to all aspects of breast cancer research and prevention, from corporate profits to environmental contaminants, pharmaceuticals, "pink"-washing,[7] and individual breast cancer sufferers and survivors. Building on Rachel Carson's work (1962) uncovering the links between environmental chemicals and their impact on birds, other animal species, and ecosystems, feminist environmentalists exposed the links between synthetic chemicals and the endocrine systems of human and nonhuman animals. From pesticides and plastics to paint and pajamas, synthetic chemicals are linked to the feminization of male reproductive systems in frogs and other wildlife (Aviv 2014) and associated with breast cancer in women. Lois Gibbs' work on dioxin (1995), Liane Clorfene-Casten's work on breast cancer (1996; 2002), Theo Colborn's exposé of synthetic chemicals (1997), and Sandra Steingraber's (1997; 2001) eloquent studies of agricultural chemicals, environmental health, children's health, and human cancers are all landmark contributions to our understanding of the interconnections among environmental health, public health, and social justice. This feminist health and environmental science research has contributed to the scientific and epidemiological foundations of the environmental justice movement, and provides longstanding environmental feminist foundations for material feminist theorizing.

A second branch of feminist theory, feminist animal studies has explored the links between the production, transport, consumption and waste of animals used in industrial food systems, and that industry's many assaults on human and environmental health. Today's industrialized production of animal bodies for human consumption emerges from a constellation of oppressive practices. Building on earlier feminist research into the exploitation of female reproduction (Corea 1985), and the development of reproductive technologies via experimentation on nonhuman females first, feminist animal studies scholars have emphasized how Western systems of industrial animal production ("factory farming") rely specifically on the exploitation of the female (Adams and Donovan 1995; Donovan and Adams 2007), harming the health of both nonhuman females and the human females who consume their bodies and their reproductive "products." As Carol Adams (2003) points out, "to control fertility one must have absolute access to the female of the species" (147). The control of female fertility for food production and human reproduction alike uses invasive technologies to manipulate female bodies across the species (Adams 2003; Corea 1985; Diamond 1994):

- Battery chickens are crowded into tiny cages, de-beaked, and inoculated with numerous antibiotics to maximize control of their reproductive output, eggs (Davis 1995). Male chicks are routinely discarded because they are of no use to the battery hen industry, while female chicks are bred to deformity with excessively large breasts and tiny feet, growing up to live a radically shortened lifetime of captivity, unable to perform any of their natural functions (i.e., dustbathing, nesting, flying).
- Pregnant sows are confined to gestation crates and after they give birth they are allowed to suckle their offspring only through metal bars.
- Dairy cows are forcibly inseminated, and their male calves are taken from them twenty-four to forty-eight hours after birth and confined in crates, where they will be fed an iron-deprived diet until they are slaughtered for veal.[8]

Bridging affect theory and feminist animal studies, Lori Gruen (2012) proposes the concept of "entangled empathy" as a strategy for reminding humans of our intra-actions across species and food production systems. Entangled empathy is an affect co-arising with our recognition of the affective states of other beings; its energetic and embodied awareness motivates action to eliminate suffering.

Describing animals used in these industrial food systems as "workers" (Haraway 2003) is reprehensible for the ways that it obscures the institutionalized oppression of reproductive labor and human responsibility, as Zipporah Weisberg (2009) explains, for who would choose a "job" requiring a lifetime of imprisonment, separation from one's family, the murder of one's offspring, along with crowding, biological manipulation to the point of crippling, all culminating in execution? In her work "bringing together environmental, climate and reproductive justice," Giovanna Di Chiro (2009) defines reproductive justice as involving not just "bodily self-determination and the right to safe contraception" but also "the right to have children and to be able to raise them in nurturing, healthy, and safe environments" that requires an availability of "good jobs and economic security, freedom from domestic violence and forced sterilization, affordable healthcare, educational opportunities, decent housing, and access to clean and healthy neighborhoods" (2). Linking the exploitation of sexuality and reproduction across species as a feature of the colonialist and techno-science worldview, feminist animal studies scholars have described industrial animal food production as a failure of *reproductive, transspecies, and environmental justice.*

It's also a matter of climate justice, as the UN Food and Agricultural Organization Report "Livestock's Long Shadow" (2006) confirms. The report defines "livestock" as all animal foods, including cattle, buffalo, small ruminants, camels, horses, pigs, and poultry; livestock products include meats, eggs, milk and dairy. The "factory farming" first introduced in the United States has been exported globally, to the detriment of the planet. Increasing areas of cropland are being used to feed cattle and other food animals; forests are being replaced with rangeland; vast quantities of water are used to irrigate crops for food animals and given to food animals for drinking. The wastes of industrial animal food production—which includes pesticides, herbicides, fertilizers, hormones and antibiotics, manure, and the wastes from slaughterhouses—contaminate wetlands and wildlands, and have produced the hypoxic ("dead zone") area at the Mississippi River's outflow in the Gulf of Mexico. Methane produced by flatulence, carbon dioxide produced through respiration and transport, nitrous oxide and ammonia are all GHGs multiplied through industrial animal agriculture. Livestock production not only exponentially increases our planet's GHG emissions, it also reduces the GHG-absorbing areas of forests, the "carbon sinks" whereby the planet might restore a balance.

Human health is also variously affected. Meat production is associated with prosperity, good health, social status, and the affluent lifestyle of the Western industrialized countries. As more and more nations seek to emulate the meat consumption levels of the industrialized world, their rates of cancer, heart disease, obesity, and other animal food-related illnesses increase (Campbell & Campbell 2006). Statistics comparing the growing obesity of first world overconsumers and two-thirds world persons suffering from hunger and malnourishment can be correlated with the rates of animal food consumption, and with the gendered character of hunger (FAO 2013). In developing countries, women account for 43% of the agricultural labor force, although their yields are 20–30% lower than men's because women are barred from farming the best soils, and denied access to seeds, fertilizers, and equipment (WFP 2013). Around the world, it is women who are responsible for cooking and serving food, and it is men who eat the first and most nutritious foods, leaving children to eat afterwards, and women to eat last. When there is insufficient food, women deny themselves food so that children can eat: while an estimated 146 million children in developing countries are underweight due to acute or chronic malnutrition, 60% of the world's hungriest are women (WFP 2013). According to the World Food Program, if women farmers had the same access to resources as the men do, the number of hungry people in the world could be reduced by up to 150 million (WFP 2013).

Industrial animal food production has been described as "a protein factory in reverse" (Robbins 1987), largely because eating high on the food chain requires more "inputs" of grain, water, and grazing land. The ecological and human toll of industrialized animal agriculture is no longer debated, for the facts are well known:

- It takes 13 pounds of grain and 2,400 gallons of water to produce 1 pound of meat, and 11 times as many fossil fuels to produce one calorie of animal protein versus plant protein.
- Raising animals for food requires 30% of the earth's surface.
- There is currently enough food in the world to feed approximately 12 billion people, yet over 900 million are hungry (UNFAO 2006; WFP 2013).

As food and development scholars have argued for decades, hunger is not a problem of overpopulation but rather one of distribution, and elite control of the world's food supply (George 1976, 1984; Hartman and Boyce 1979; Lappé and Collins 1998). Moreover, debt repayment programs (called "structural adjustment") require developing countries to produce cash crops for export rather than food crops for subsistence as a way to pay off debt; biotechnology corporations promote high-yield seeds which require expensive inputs of fertilizer and monocropping techniques that displace subsistence foods, destroy biodiversity, and lower water quality, producing both debt and hunger. These facts notwithstanding, the worldwide production of meat and dairy is projected to more than double by 2050 (UNFAO 2006). Industrialized animal food production is simultaneously a problem of species justice, environmental justice, reproductive justice and food justice. For too long, "food justice" has been defined solely in terms of justice across human diversities, but *authentic food justice cannot be practiced while simultaneously excluding those who count as "food."* Food justice requires interspecies justice, which intersects with reproductive justice and queer justice alike.

Queer food justice grows out of today's budding ecoqueer movement, which Joshua Sbicca (2012) defines as a "loose-knit, often decentralized set of political and social activists" who challenge the dominant discourses of sexuality, gender, and nature as a means for deconstructing hegemonic knowledge systems (33–34). Reviewing the herstory of queer ecoactivism in building lesbian ecocommunities and music festivals, and in challenging the heteronormativity of urban parks through gay cruising and public sex (Mortimer-Sandilands and Erickson 2010), Sbicca focuses particularly on the queer food justice movement being shaped by queer farmers and gardeners who may not feel comfortable in the alternative food movement, whose most visible U.S. representatives—Michael Pollan, Eric Schlosser, Joel Salatin, Barbara King-solver—are largely white, heteromale, and middle class. The grassroots food justice movement is far from this stereotype, and reaches back to European women's gardens of the eighteenth century (Norwood 1993), Black women rural gardeners in the post-Reconstruction South (Walker 1983), and women rooftop gardeners in Harlem. Formed in 2007, San Francisco's Queer Food For Love (QFFL) seems like a queer update of Food Not Bombs with their desire to provide food, community, and a safe space against prejudice. Similarly, San Francisco's Rainbow Chard Alliance, formed in 2008, bridges the organic farming movement and the queer movement, creating community for like-minded "eco-homos" in the Bay Area and California (Sbicca 2012). Not confined to California, the queer food justice movement is articulated through groups ranging from Vermont, Massachusetts, and Connecticut to Tennessee, Alabama, Arkansas, Kansas, and Washington. Concerned about the intersections between environment, sexuality, and gender, these queer groups use food to build community, fight oppression, and take care of planetary and human bodies, though it's not clear whether these groups make connections between sexuality and species oppressions, and thus enact a species-inclusive food justice as well.

With these facts of world hunger, food production, gender, sexuality and species restored to an analysis of climate change, charging human overpopulation as a root cause of climate change seems misguided at best: instead, climate change may be described as *white industrial-capitalist heteromale supremacy on overdrive*, boosted by widespread injustices of gender and race, sexuality, and species. Eating high on the food chain must be seen as tilting the planet's plate of food into the mouths of the world's most affluent, at the cost of billions of nonhuman animal lives, and up to 870 million people—almost half of them children under the age of 5—who suffer from chronic undernourishment (FAO 2013). Population control and industrialized animal food production are no substitute for reproductive justice, interspecies justice, gender justice, and climate justice.

WHAT DO WE WANT? A MORE INCLUSIVE CLIMATE JUSTICE

The 27 Bali Principles of Climate Justice (2002) redefine climate change from an environmental justice standpoint, using as a template the original 17 Principles of Environmental Justice (1991) created at the First National People of Color Environmental Summit. The Bali Principles address the categories of gender, indigeneity, age, ability, wealth, and health; they provide mandates for sustainability in energy and food production, democratic decision-making, ecological economics, gender justice, and economic reparations to include support for adaptation and mitigation of climate change impacts on the world's most vulnerable populations. These principles restore many of the missing components of climate science's "truncated narrative" (Kheel

1993), connecting the unsustainable consumption and production practices of the industrialized North/First World (and the elites of the South/Two-Thirds world) with the environmental impacts felt most harshly by those in the South and the impoverished areas of the North. Yet, despite their introductory Principle 1 "affirming the sacredness of Mother Earth, ecological unity and the interdependence of all species," the Bali Principles are not informed by a posthumanist perspective. Just as "Climate Justice affirms the need for solutions that address women's rights" (Principle 22), climate justice also needs to affirm solutions that address queer rights; just as "Climate Justice . . . is opposed to the commodification of nature and its resources" (Principle 18), climate justice also needs to oppose the commodification of animal bodies and female bodies across species. To be inclusive, the Bali Principles need to be augmented with a queer, feminist, and posthumanist justice perspective.

On November 12, 2013, an unprecedented workshop on gender balance and gender equality was held at the UNFCCC's COP19 in Warsaw, Poland, where for three hours, speaker after speaker disclosed facts confirming women's marginalization from climate change decision-making: "the number of all women participating as delegates in UNFCCC processes, or as members of constituted bodies still falls below 35%, and as low as 11–13% in the case of some constituted bodies" (GGCA 2013). A list of eight solutions proposed by the panelists included basic affirmative action strategies complete with quotas, sanctions, and a monitoring body to keep track of gender balance; funding for participation and training; and tools and methodology to guide research and practices promoting "systematic inclusion of women and gender-sensitive climate policy" (GGCA 2013). These changes enacting gender equity provide a necessary first step toward a more transformative feminist analysis and response to climate change. That it has taken more than two decades since WEDO's "Agenda 21" for this workshop to occur offers visible confirmation of the masculinist character of climate change analyses—and the dedicated persistence of women drawing on liberal and cultural feminist strategies.

But, does bringing women more fully into the United Nations' discussions on climate change promise to bring forward a feminist perspective? Scholars have investigated whether women's representation in decision-making bodies affects environmental outcomes (Ergas & York 2012), whether a higher participation of women leads to better climate policy (Alber & Roehr 2006), and whether there is any verifiable gender difference in climate change knowledge and concern (Alaimo 2009; McCright 2010).[9] Summarized in Table 6.2, the data suggest that women would act differently than men in decision-making positions about climate change problems and solutions. Yet at least one source (Rohr 2012) cites an exception in the Commissioner on Climate Action, Connie Hedegaard, who was "not in favour of addressing

Table 6.2   Gender Differences in Climate Change Knowledge, Attitudes, Actions

- Women are estimated to compose between 60% and 80% of grassroots environmental organization membership, and are more active in environmental reform projects*
- Women tend to perceive environmental risks as more threatening* and express greater concern about climate change than do men***
- Women in the United States show greater scientific knowledge of climate change***, approach the issue of climate change differently, and express different concerns and potential solutions to problems*
- Women consider climate change impacts to be more severe**
- Women are more skeptical about the effectiveness of current climate change policies in solving the problem, whereas men tend to put their trust in scientific and technical solutions**
- Women are more willing to change to a more climate-friendly lifestyle**
- Climate protection policy areas—energy policy, transportation planning, urban planning—tend to be male dominated**
- Women are underrepresented in areas of climate change policy**
- Women underestimate their climate change knowledge more than do men***

Data *Sources*: *Ergas & York 2012; **Albert & Roehr 2006; ***McCright 2010.

gender in European climate policy, because she deemed it relevant only for developing countries" and didn't want to be "overloaded" by integrating gender aspects (2). Thus, while gender balance at all levels of climate change decision-making is necessary, it "does not automatically guarantee gender responsive climate policy" (Rohr 2012, 2). A wider transformation is needed, involving "progressive men [and genderqueer others] who are prepared to question their masculinity and gender roles," and work together to uncover "the embedded gender [sexuality] and power relations in climate change policy and mitigation strategies" (Rohr 2012, 2). From these studies, it appears that structural gender inequality, and more specifically the underrepresentation of women in decision-making bodies on climate change, is actually inhibiting national and global action in addressing climate change.

Given the correlation and mutual reinforcement of sexism and homophobia (Pharr 1988), it should be no surprise that the standpoints on climate change for women and LGBTQ populations are comparable. Yet in UN discourse to date, when LGBTQ people seek an entry point into the ongoing climate change conversations, the primary entry point is one of illness, addressing only HIV and AIDS (McMichael, Butler and Weaver 2008). Very few studies have recognized a *queer ecological perspective* (Gaard 1997; Mortimer-Sandilands and Erickson 2010), much less brought that perspective to climate change research and data collection. Nonetheless, these few studies confirm that the link between climate change and various LGBT individuals and communities stems from "the fundamentalist desires to dominate and control other people's environment, resources, contexts and desires" (Somera 2009).

According to a U.S. poll conducted by Harris Interactive, "LGBT Americans Think, Act, Vote More Green than Others" (2009), a conclusion based on answers to several key questions about whether it is important to support environmental causes, whether climate change is actually happening right now, whether the respondent would self-identify as an environmentalist, and whether it is important to consider environmental issues when voting for a candidate, buying goods and services, or choosing a job (See Figure 6.2).[10] Most significant in the Harris Poll—given that heterosexuals are more likely to have children—was the LGBT response expressed for what kind of planet we are leaving for future generations, a question which concerned LGBT respondents at 51% as compared with 42% of heterosexual respondents. Exploring the ways that "non-white reproduction and same-sex eroticism" are constructed as "queer acts against nature" in both environmentalist and homophobic discourses, Andil Gosine (2010) sees both as "threatening to the white nation-building projects engendered through the process of colonization" (150). Discourses on the ecological dangers of overpopulation and queer sexualities are alike, Gosine argues, in that both deny the erotic (cf. Lorde 1984). The toxic environments of climate change and homophobia are linked in the reason/erotic dualism of the Master Model (Plumwood 1993), and cohere with other linked dualisms of white/non-white, wealthy/poor, intellectual/reproductive, a linkage that has been called erotophobia (Gaard 1997) and ecophobia (Estok 2009).

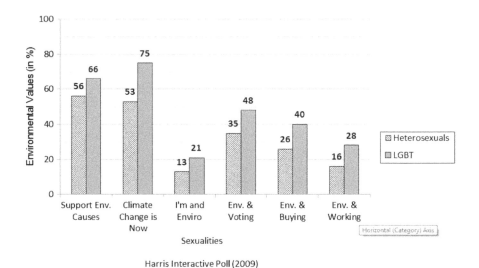

Harris Interactive Poll (2009)

Figure 6.2:   Harris Interactive. 2009. "LGBT Americans Think, Act, Vote More Green than Others." Business Wire (New York). 26 October.

The culturally constructed fear, denial, and devaluation of our embodied erotic are not lost on ecoactivist youth, who are among the first to mention sexual well-being in climate change discussions. At COP18 in Doha, Qatar, November 26–December 8, 2012, a passionate youth movement emerged, according to WEDO: "The Youth Gender Working Group emphasized issues like the right to financing and technology, how disasters impact women, LGBT communities, sexual health and reproductive rights" (De Cicco 2013). These explorations of queer feminist ecology can augment the slogan of the Gender and Climate Change Network (genderCC): "There will be no climate justice without [queer] gender justice."

HOW DO WE GET THERE? GOALS AND OUTCOMES

Implementing the Bali Principles with their queer feminist posthumanist augmentations requires transformative strategies that are both top-down and bottom-up; the responsibilities are both systemic, requiring changes in national and corporate policies, and personal, requiring changes on the part of citizens and consumers (Cuomo 2011). Some techno-science solutions to climate change can help to mitigate the outcomes of first world nations' and corporations' unjust and antiecological practices, and transform our energy reliance to more sustainable sources, but a queer feminist climate justice approach goes to the roots and calls for equity and sustainability at every level, from citizen to corporation, and it begins with economics.

As feminist economist Marilyn Waring observed in her classic work, *If Women Counted: A New Feminist Economics* (1988), the UN System of National Accounts (UNSNA) has no method of accounting for nature's own production or destruction until the products of nature enter the cash economy, nor does this system account for the majority of work done by women. A clean lake that offers women fresh water supplies for cooking and crops has no economic value until it is polluted; then companies must pay to clean it up, and the cleanup activity is performed by men and recorded as generating income. Similarly, living forests which supply women with food, fuel, and fodder have no recorded value in the UNSNA until they are logged and their products can be manufactured into commodities for sale—then all related industry and manufacture, usually seen as men's work, is recorded as income generating. In *The Price of Motherhood*, Ann Crittendon (2001) addresses the shadow economy of women's unpaid labor in reproduction and caregiving, linking the gendered economy with ecological economics.

For posthumanist accuracy and ecojustice sustainability, we need a feminist ecological accounting system, capable of tracking and promoting climate justice economic practices at every level, from local to global.

Replacing economic globalization (which in practice has meant global corporatization and indigenous as well as ecological colonialism) with global economic justice offers a frontal assault on climate change. Industrialized nations must pay our climate debts both to communities and to ecosystems, as called for in the Bali Principles, and develop economic accounting practices that do not externalize the costs of a just transition onto the environment and communities facing the outcomes of climate change. An economic transition from excessive takings (i.e., "profits") from women, indigenous communities, the two-thirds world, animals, and ecosystems to a green economy requires sustainable jobs of the kind advocated by Van Jones' organization, Green for All. These jobs will include sustainable energy systems, sustainable transit systems, and urban planning guided by environmental justice.

The foundations for food justice have been growing for decades in the food cooperative movement which began in the nineteenth century, and was more recently resurrected in the 1970s. Today's food justice movement includes Community Supported Agriculture (CSA), the advent of rooftop and community gardens exemplified by groups such as Will Allen's Growing Power in Milwaukee, queer food justice farmers and gardeners from Vermont to California, and Natasha Bowens' "Brown Girl Farming" efforts to map food justice so that the food movement is not seen as the domain of affluent consumers but is shaped by the self-determination of women and communities of color (Bowens 2013). With a posthumanist food justice movement reconceived to include other animal species and to consider their lives in terms of reproductive justice, the animal sanctuary movement—a corrective response of entangled empathy, interrupting the practices of industrialized animal agriculture—may face a new opportunity: freeing up the excessive land space now used by industrialized animal agriculture, small-scale farming and community gardens alike will have more land for farming and for freed animals. This transition away from industrial animal agriculture begins by ceasing the artificial insemination of female animals on factory farms, and possibly returning freed animals to live out their lives adjacent to community gardens and small farms, where they can provide cropping services and fertilizer, giving humans a chance to repay our interspecies debt.

Overlapping with food justice, the Transition Town movement, named in 1998 and formally launched by 2005, has spread from its origins in the United Kingdom to countries on every continent, with communities responding to peak oil by building local food security through community gardens and local energy security through renewables. Some groups build on the movement for local currencies based on barter: one hour of anyone's time is equal to another's.

As Bill McKibben wrote in his *Rolling Stone* article, "Do the Math" (2012), social and environmental movements of the kind needed now are often inspired by having an enemy. Pinpointing the globalized fossil fuel industry, McKibben launched 350.Org's strategy of divestment, modeled on the successful divestment strategies that prompted South Africa to end apartheid. Withdrawing financial support from systems destructive of global ecojustice is another necessary but not sufficient method of resistance. While crucial to a just transition, economic boycotts and microlevel community infrastructures providing an alternative to global capitalism through local economics, energy, food, and governance can still be overridden by global-level trade agreements, multinational investments, and other forms of economic or militarized pressure. Withdrawing economic support from these global institutions of ecological domination, investing in systems based on social/environmental/climate justice, and pressuring for equitable representation within the international institutions of governance are equally crucial strategies.[11]

The macro-level discussions at the UNFCC must be gender balanced, as was suggested over twenty years ago by the WEDO in their 1991 Preparatory Conference for the UNCED in Rio de Janeiro, 1992. There, many of the most salient issues of climate change were both addressed and ignored in these two pivotal conferences (Brú Bister and Cabo 2004). What feminist climate justice scholars also note, albeit as an afterthought, is that these discussions of "gender and climate" have tended to focus only on women. More research is needed on the ways that men around the world have variously benefited from or been affected by climate change discussions, problems, and outcomes. More research is needed on the gender roles of masculinities in diverse cultures, and the ways these social constructions promote overconsumption, sexual violence and exploitation, the abandonment of family members during climate change crises, and rationalize the de facto exclusion of women from decision-making bodies at the local, national, and global levels. Much has been written confirming the antiecological construction of masculinity (Kheel 2008). It is time to envision and to recuperate culturally specific, ecological masculinities that will companion this transition to climate justice (Gaard 2014), and in this regard, posthumanist genderqueer activists will have much to offer.[12]

TOWARD AN ECOFEMINIST CLIMATE JUSTICE

Feminist scholars have invoked the concept of intersectionality (Crenshaw 1991; Collins 1990) in order to describe the "intra-actions" (Barad 2007) of race, class, gender, sexuality, ethnicity, age, ability, and other forms of human difference, using this analysis to develop more nuanced understandings of power, privilege, and oppression. But fewer scholars have critiqued the humanism of intersectionality (Lykke 2009), or proposed examining the

exclusions of species and ecosystems from intersectional identities, addressing the ways that even the most marginalized of humans may participate in
the Master Model process of instrumentalisation when it comes to nonhuman
nature and *earth others*. As an ecological identity and ecopolitical standpoint
resisting the Master Model, ecofeminists once proposed the self-identity of
"political *animal*" for first world ecocitizens (Gaard 1998; Sandilands 1994,
1999); this view resituates humans within ecosystems and faces us toward
assessing ecosystems flows and equilibrium, while simultaneously attending
to the well-being of our transcorporeality (Alaimo 2008).[13] Joining a philosophical reconception of human identity with an ecopolitical exploration of
economic globalization and its role in producing climate change, a queer
posthumanist and "feminist ecological citizenship" (MacGregor 2014) could
send a critical challenge to the techno-science discourse about "mitigation
and adaptation" (rather than reduction and prevention) currently dominating
responses to climate change (i.e., geoengineering).

How much more time do we have to lose?

NOTES

1. Although I was sitting with three other queer ecofeminist women at the WEDO
conference, none of us thought to theorize the connections between our sexualities
and climate change; the historical moment prompted us to challenge the essentialist
rhetoric of "mothering earth," and to focus on intersections of race, gender, species,
ecology, democracy, and economic globalization, laying the groundwork for future
studies. Noël Sturgeon discusses my disappointment with the WEDO 1991 conference in her *Ecofeminist Natures: Gender, Feminist Theory and Political Action* (New
York: Routledge, 1997), 159.

2. Chandra Talpade Mohanty (2002) discusses the terms *Western/Third World*,
North/South, and *One-Third/Two-Thirds Worlds* as different ways of approaching
descriptions of differences in affluence, power, and the history of colonization (506–
508). She acknowledges that all of these terms are imprecise, and resorts to using
some terms in combination (i.e., *First World/North*, *Third World/South*). Confronting
the same problems in searching for sufficiently precise terms, I will simply follow her
lead.

3. These statistics are widely cited by international NGOs; see, for example, UN
Women.Org, UNICEF, Millennium Campaign/Voices Against Poverty, and more. It
seems there are multiple sources confirming these statistics, first cited in 2007, and
unchanged in 2014.

4. The gendered impacts of climate change on women in the Two-Thirds World
as discussed in this paragraph appear in numerous sources; see Dankelman 2010;
Duncan 2008. The presentation of women as vulnerable victims of climate change is

both cited and strongly critiqued in MacGregor 2010, Resurrección 2013, and Tuana 2013.

5. I develop a theory of queer climate justice in my chapter, "Out of the Closets and into the Climate!" (2017).

6. See http://conceivablefuture.org/FAQ (accessed 11/24/2016).

7. The "Think Before You Pink" campaign challenges the pink ribbons associated with the many fundraising races and marches to "end" breast cancer; the funds go toward cancer researchers, not environmental toxicologists, and certainly not toward implementing the Precautionary Principle which would prevent industrial chemicals from being sold until tests had proven conclusively that the chemical posed no harm to humans, animals, or ecosystems. See Liane Clorfene-Casten (2002) and Breast Cancer Action's "Think Before You Pink" website.

8. The horrendous suffering caused by industrial animal agriculture is widely documented in books and internet sources by Farm Sanctuary, PETA, Vegan Outreach, Mercy for Animals, the American Society for the Prevention of Cruelty to Animals, the Humane Society of the United States, Sustainable Table, and more. Notable recent publications include Paul Solotaroff's exposé, "In The Belly of The Beast" for *Rolling Stone* (December 10, 2013) and the 2014 release of "Cowspiracy: The Sustainability Secret" (Anderson & Kuhn), a documentary revealing the complicity of mainstream environmental organizations in covering up the climate change-factory farming links.

9. The careful methodology of these studies affirms their validity. International findings on gendered differences in climate change causes, analyses, and solutions in Ergas & York (2012) rest on sixty peer-reviewed studies, which then shape the questions and statistical analysis these authors undertake. McCright (2010) tests the arguments about gender differences in scientific knowledge and environmental concern using eight years of Gallup data on climate change knowledge and concern in the U.S. public. Alber & Roehr (2006) report on the project "Climate for Change—Gender Equality and Climate Policy" that performed data surveys of the gender balance in climate policy at local and national levels for ten major cities in four European countries (Germany, Italy, Finland, Sweden).

10. Because the findings may surprise some readers, I include links to Harris Interactive Methods for LGBT surveys: http://www.harrisinteractive.com/MethodsTools/DataCollection/SpecialtyPanelsPanelDevelopment/LGBTPanel.aspx.

11. The socially responsible investing movement has eighteenth-century roots in religious communities of Quakers and Methodists, with its values revived and augmented by twentieth century social movements for civil rights, workers' rights, peace and environmental health. The Occupy Movement launched in September 2011 drew on the strategies of socially responsible investing in its "Move Your Money" or "Ditch Your Bank!" campaign, urging social justice-minded citizens to divest from corporate banks and invest in credit unions and community banks. A queer feminist and posthumanist discussion of socially responsible investing is long overdue. See http://www.ussif.org/ for this movement's most recent articulation as "sustainable and responsible" investing.

12. See chapter 8 for a discussion of Beth Stephens and Annie Sprinkle's "Goodbye Gauley Mountain: An Ecosexual Love Story," a documentary bringing the tools of queer sexuality and erotic love for the earth to support communities in West Virginia's Appalachian mountains as they fight coal mining, mountaintop removal, and the related harms to human, animal, and environmental health.

13. For nonwestern, indigenous communities, the "indigenous cosmopolitics" described by Marisol de la Cadena (2010) may be a better fit. My theorizing applies primarily to my own cultural and economic contexts in a first world industrialized nation.

Chapter 7

"Cli-Fi" Narratives

In the month that I begin to write about climate change, my body begins to sweat.

I wake in the night, flooded with heat. The warmth begins at the base of my skull, then curls up my head and around my neck like smoke curling under a closed door. The heat pours down my neck and shoulders, arms, spine, leaving me sweating, then chilled.

Nothing is wrong. This sudden heat is a step toward my own mortality, natural and inevitable.

I think about what it is like to be a body, overheating.[1]

How can feminists and ecocritics empathize with *earth others* and use that empathy to motivate our scholarship? To date—eschewing both empathy and empirical data—the United States business, government, and media have resisted the dire warnings echoing from environmentalists to politicians, NASA scientists, and the United Nations' IPCC, a group of over 300 scientists and government officials whose reports in 1990, 1995, 2001, 2007, and 2013 confirm the fact that anthropogenic (human-created) GHG emissions could, if left unchecked, raise global average temperatures by as much as 5.8 degrees Celsius (or 10.4 degrees Fahrenheit) by the end of this century. From the grassroots, Idle No More indigenous activists, Occupy activists, forest-dwellers in Brazil, women farmers and scientists in India, along with citizens and leaders of low-lying nations all affirm the urgency of global climate change—its immediate effects as a humanitarian and ecological crisis—as possibly the most pressing issue on the international environmental agenda. Yet first world citizen-consumers have been slow to listen, and slow to demand institutional

---

1. These epigraphs record my journal notes, written in the margins beside my more scholarly argument. I include both voices here in this chapter.

changes, lulled into complacency, in part, by propaganda from the mainstream media, and the half-truths of climate change science fiction ("cli-fi"). This chapter offers an empathic, empirical, and embodied intervention.

Given the proven power of narrative to shape public opinion and to mobilize social movements,[1] I propose bringing a critical ecofeminist standpoint—bridging feminist animal studies, posthumanism, material feminism, ecofeminism, and feminist ecocriticism—to illuminate the strengths and shortcomings of literary narratives that present the problem of climate change from a (masculinist) technological-scientific perspective. Provided with a more complete restor(y)ing of climate change causes and effects, authors and readers of literature, as well as activists and ecocritics will have a greater potential to shape and contribute to activist and policy-making discourses around climate justice.

Ecocritics have already observed the disjunction between the environmental sciences and the environmental humanities (Buell 2005; Garrard 2004), noting the dominance of environmental sciences in defining environmental problems and controlling the discourse around their solutions. These enviroscience analyses offer incomplete descriptions without the perspective of the environmental humanities: fields such as ecopsychology, public health, environmental philosophy, environmental politics, environmental economics, and ecocriticism provide critical information that augments and often transforms our understanding of environmental problems—particularly in the case of climate change. To begin, I explore a variety of climate change narratives that fail to challenge the underpinnings of colonialism, neoliberalism, speciesism, and gendered fundamentalisms.[2] By expanding the genres and geographies of ecocritical analysis to include artists of color and of diverse sexualities, and addressing in narratives and analyses the climate-changing practices of animal food production and consumption, a critical ecofeminist perspective may offer a more inclusive restor(y)ing of climate change narratives.

CLIMATE CHANGE FICTION AND "CLI-FI" SCIENCE FICTION

Despite a wealth of feminist environmental writing from the 1980s to the present, ranging from environmental justice-themed science fiction by Octavia Butler (i.e., *The Xenogenesis Trilogy, Parable of the Sower*), Ursula LeGuin (*The Word for World Is Forest, Always Coming Home, Buffalo Gals and Other Animal Presences*), and Marge Piercy (*Woman on the Edge of Time*) to futurist apocalyptic and postapocalyptic novels such as Starhawk's *The Fifth Sacred Thing*, Cormac McCarthy's *The Road*, and Margaret Atwood's *MaddAddam* trilogy, the feminist fiction confronting climate change has yet to be written.[3] To date, the prominent texts of climate change fiction and science fiction discussed herein are largely male-authored and nonfeminist at best, or antifeminist and sexist at worst—problematic in that feminist research has traced the root causes of climate change to humanist, colonialist, antidemocratic, and antiecological beliefs and practices (MacGregor 2010).

In climate change fiction, the most literary representative of the genre is T. C. Boyle's *A Friend of the Earth* (2000), its title a reference to the U.S.-based international organization, Friends of the Earth, founded in 1969 by David Brower. Set somewhere around Santa Barbara, California, in the year 2025, this fictional narrative suggests that global warming is a consequence of economic, cultural, and political forces that have produced unsustainable population growth, irreversible loss of biodiversity, deforestation, species extinction, and an end to social supports such as health care and social security. Although the narrative ends on a comedic note, promising recreation of the heteronuclear family and thus the perpetuation of the human species, the overwhelming tone of *A Friend of the Earth* is one of cynicism and despair: its narrative solution is withdrawal from society, since civic engagement hasn't worked. Boyle's analysis of global warming includes the ecological, social, cultural, economic, and political causes and consequences, though with a focus on the white middle class; the book omits discussion of diversity such as gender, race, sexuality, and nation when addressing climate change problems or solutions. But climate change is not an equal opportunity disaster, undifferentiated in its impacts on diverse nations, communities, genders.

In climate change science fiction (first called "cli-fi" in 2007, by journalist Dan Bloom), the message is sometimes reversed: according to skeptic Michael Crichton's *State of Fear* (2004), climate change is a hoax produced by environmentalists so determined to promote fear of climate change that they use exotic technologies to start natural disasters (crumbling a massive Arctic glacier, triggering a tsunami), and are willing to see innocent people die, just to make their case. Crichton's narrative portrays the "experts"— which include PhDs, scientists, intellectuals, and feminists—as spectacularly corrupt and terribly wrong. His heroic skeptic, John Kenner, is companioned by a trusty Nepalese sidekick, Sanjong, echoing the racist and not-too-subtle homoerotic pairing of John Wayne and Tonto from U.S. Westerns of the 1940s that celebrated the epic myth of Euro-American colonialism. In Crichton's novel, the pair work together to provide charts, graphs, and other "hard" data to disprove global warming. By the novel's conclusion, one environmentalist has been fed to cannibals, and the skeptics have become suddenly irresistible to women, all in nine days. If the narrative itself isn't sufficiently alarming, ecocritics may find the book's popularity with the uncritical and unscientific public even more disturbing, and appalled by its use as a textbook for the honors seminar, "Scientific Inquiry: Case Studies in Science" at SUNY-Buffalo.[4]

Kim Stanley Robinson's climate change trilogy—*Forty Signs of Rain* (2004), *Fifty Degrees Below* (2005), *Sixty Days and Counting* (2007)—is a welcome counterpoint to Crichton's polemic, at least from an ecocritical standpoint. The first text in the series sets up the problem of global warming when Washington, DC, is hit hard with two days of rain, portions of the city are flooded, and animals are released from zoos so they don't drown. The second text shows the more developed consequences of global warming: the Gulf Stream has stalled, causing frigid winter temperatures in the Eastern United States and Western Europe. As people starve, multinational corporations find ways of making a profit (which is the same plot as in Nathaniel Rich's *Odds Against Tomorrow*). Antarctica's ice shelves collapse, and low-lying nations sink under the waters. In Washington, DC, environmental scientists must overcome government inertia to put in place policies that may save the world.

While the second novel focuses on the market failures of capitalism and democracy in the west, the third novel's narrative suggests that the world would be better if scientists took over politics. Scientists fill the White House, but the book does not explore the benefits and drawbacks of a scientocracy (science's false claims of objectivity and its universalizing tendencies, as well as the corporate control of science being chief among them), and seems to promote the idea that everything would be better if only the right *man* were President. Indeed, all of Robinson's main plots revolve around men, the main characters are men, and the proposal that a male-centered ecosocialist scientocracy will solve problems of climate change without addressing problems of social injustice, not to mention simple gender parity, seems limited at best.

A spate of "cli-fi" novels has appeared in the years 2011–2013, effectively defining this new genre. Novels which take climate change as a fact and set their plot around characters who must deal with the fallout include Paolo Bacigalupi's *The Windup Girl* (2009), Ian McEwan's *Solar* (2010), Barbara Kingsolver's *Flight Behavior* (2012), Daniel Kramb's *From Here* (2012) and Nathaniel Rich's *Odds Against Tomorrow* (2013). A review essay by Rebecca Tuhus-Dubrow (2013) discusses seven of these novels, and a website on "Cli-Fi Books" run by a British Columbia micropress, Moon Willow, was launched in August 2013, and renamed simply "Eco-Fiction" by 2015 (Woodbury). A reviewer for the *New York Times*, Richard Pérez-Peña (2014) notes that cli-fi fits "a long tradition of speculative fiction that pictures the future *after* assorted catastrophes" and omits analysis of the catastrophe's root causes and solutions: cli-fi addresses adaptation and survival, but "if the authors are aiming for political consciousness-raising, the effort is more veiled than in novels of earlier times." Recent feminist ecocritical analyses of climate change narratives concur with Pérez-Peña. For example, Christa Grewe-Volpp's (2013) discussion of Octavia Butler's *Parable of the Sower* (1993) and Cormac McCarthy's *The Road* (2006) as climate change postapocalyptic novels which depict characters struggling to survive in degraded environments, and Katie Hogan's (2013) queer green analysis of climate change references in Tony Kushner's apocalyptic play, *Angels in America* (1994), linking illnesses that are simultaneously social, ecological, and psychological. As this ecocritical scholarship shows, the focus of cli-fi narratives remains confined within the apocalyptic failure of techno-science solutions, and uninformed by the global climate justice movement.[5] After reading such narratives, cli-fi readers take home the message that climate change is a failure of technology and science, not a failure of species justice or environmental justice, and thus their actions after reading these books might include renewed faith in techno-science solutions and individual carbon-footprint reduction, rather than working for system-wide ecojustice changes.

CLIMATE CHANGE NONFICTION WRITING

In the field of literary nonfiction, the majority of texts provide narratives that make environmental sciences more accessible through the lens of environmental literature, yet offer little information about environmental politics, sociology, climate justice, or ecosocial strategies for response. Two examples of this approach include Tim Flannery's *The Weather Makers* and Elizabeth Kolbert's *Field Notes from a Catastrophe: Man, Nature, and Climate Change*. Flannery's volume contains thirty-six short essays on the consequences of global warming, and in the final third of the book, he poses solutions that involve individual, national, and international actions to reduce carbon dioxide. The core of his message explores how we can shift from fossil fuels to a hydrogen-based economy, and while he acknowledges that the U.S. administration has been influenced by coal-industry donations to the Republican Party, thereby undermining political action, Flannery's environmental science solutions obscure the powerful influence of environmental economics, politics, and culture.

In refreshing contrast, Elizabeth Kolbert's work returns to the view that human politics are at the core of our responses to climate change. Her essays provide international snapshots of how global warming is affecting people, places, and species. She interviews scientists and skeptics, bringing scientific data to a humanities audience, and exposing the fallacies of global warming skeptics. Yet her conclusion to the final chapter on the "Anthropocene" (a geological epoch) offers no solutions but despair: "It may seem impossible to imagine that a technologically advanced society could choose, in essence, to destroy itself, but that is what we are now in the process of doing" (187). While there is a wealth of nonfiction handbooks countering Kolbert's despair with suggestions for "what YOU can do to stop global warming," these too are limited by their focus on individual actions in the absence of environmental context: they fail to address strategies for countering the power of multinational corporations overriding democratic decisions at the level of community, state, and nation.[6]

CHILDREN'S AND YOUNG ADULT CLIMATE CHANGE NARRATIVES

Children's environmental literature has tremendous potential for communicating messages about ecosocial justice, community empowerment, and strategies for ecodefense (Gaard 2009). But as of 2014, children's climate change literature has not realized this potential. Several texts focus on climate change effects in the Arctic (Bergen 2008; Rockwell 2006; Tara 2007), using polar bears or penguins as protagonists, and building on children's cross-species empathy to instill awareness. The solutions offered range from empathy to action, yet they articulate only environmental science's approach to climate change (i.e., switch energy sources, plant trees, bicycle, reduce consumption, and "write representatives in Congress"—but the letters' content is unspecified). And there is an ecoskeptic presence in children's literature as well. Holly Fretwell's *The Sky's Not Falling: Why It's OK to Chill about Global Warming* (2007) assures children that human ingenuity combined with an "enviropreneurial" spirit will lead to a bright environmental future, not one where people ruin the earth.

Children's films with climate change themes have not fared much better. Both "Happy Feet" (2006) and "Wall-E" (2008) use the narrative trajectory of heterosexual romance to tell stories framed by the consequences of climate change. In "Happy Feet," climate change and its root cause, elite humans' overconsumption of nature, manifest not only through ice cracking but also through the problem of human overfishing, and the resultant scarcity of fish and increase of garbage in the Arctic. In "Wall-E" the earth is completely covered with garbage, and the romance between robots Wall-E and Eve begins when a human spaceship sends a probe to see if earth can be reinhabited by the refugee population of humans who have become obese chair-bound consumers, ruled and pacified by a single corporation. In both narratives, children are invited to identify with childlike and disempowered male heroes who succeed in ecodefense and heterosexuality alike. In both films, the human change of consciousness is magical—the penguin simply confronts the overfishing, garbage-throwing humans with the plastic ring from a six-pack; Wall-E befriends the chair-bound consumers' obese leader, who "speaks truth to power" and inspires the populace to return to earth. In fiction, simply learning the facts about environmental devastation is sufficient to inspire action;

in reality, rationalism is rhetorically insufficient, climate change facts are distorted by narrative manipulation and a masculinist techno-science framework, and planetary elites (sometimes including ecocritics and our readers) are invested in current colonialist and neoliberal economic benefits and thus seem apathetic or reluctant to change.

In young adult (YA) fiction, there's been a "boom in dystopian fiction" with postapocalyptic scenarios, prompted by the immense popularity of Suzanne Collins' *The Hunger Games* trilogy (2008, 2009, 2010), whose clear cli-fi backdrop was stripped from the books' adaptation into major motion picture (2012). Writing in *The New Yorker*, reviewer Laura Miller (2010) argues that "dystopian fiction may be the only genre written for children that's routinely less didactic than its adult counterpart. It's not about persuading the reader to stop something terrible from happening—it's about what's happening, right this minute, in the stormy psyche of the adolescent reader." Evidence supporting this view can be found in Mindy McGinnis's *Not a Drop to Drink* (2010) and Saci Lloyd's *The Carbon Diaries 2015* (2013), both of which present sixteen-year-old protagonists struggling to survive environmental degradation, sociopolitical breakdown, and hormonal transitions of adolescence at the same time. Whether it is mandated water rationing (McGinnis) or carbon rationing (Lloyd), both books offer female survivalist characters reminiscent of Jean Hegland's prescient novel *Into the Forest* (1998), narrating the survivalist struggles of two teenage sisters after the collapse of civilization, by which they have lost electricity, telephones, mail, automobile fuel, commodified foods, and their parents.

Disagreeing with The New Yorker's Laura Miller, *New America Foundation* essayist Torie Bosch argues "there is a strong didactic element" in dystopian YA fiction, as "young people today are constantly told that their good behavior can—must—make up for the environmental sins of their forefathers." YA novels such as Caragh O'Brien's *Birthmarked* (2010) and Lauren Oliver's *Delirium* (2011) as well as *The Hunger Games* "are all about empowering teenagers, especially girls, to speak up and act against injustice" (Bosch 2012). In this thread, a very promising novel by Ann Brashares, *The Here and Now* (2014) depicts a time-traveling teenage protagonist, seventeen-year-old Prenna James, whose "back-from-the-future" travels with her mother and a few hundred others transport them from the climate change ravages of 2090 back to New York city in 2010, where the environmental crises of the future might still be averted. The narrative makes frank observations of a community living in "a paradise they are unwilling to relinquish even if that means dooming the future" and raises the question, "if a terrible future can be avoided, isn't it morally right to try?" (Clare 2014). Linking climate change with first world overconsumption, Brashares' *The Here and Now* seems many steps closer toward providing an accurate portrayal of climate change causes and interventions, and offers YA readers narrative inspiration for environmental activism.

CLIMATE CHANGE DOCUMENTARY AND FILM

In Al Gore's *An Inconvenient Truth* (2006), a narrative synthesis of rationalism and empathy succeeded in bringing the topic of climate change to a popular audience. The film's impact can be seen in the fact that President Bush mentioned climate change in both his 2007 and 2008 State of the Union addresses, but not in 2006; and an Internet search for the term "global warming" yielded only 129 articles for 2005, the year before the film's release, but 471 articles for 2007, the year after the film was produced (Johnson, 2009: 44). Laura Johnson attributes Gore's popular success to his capacity for moderating apocalyptic rhetoric with scientific rationalism and constructions of audience agency: at the same time that Gore gestures toward present and future climate change disasters, he simultaneously endorses new technologies and political activism. His message offers no images of either the global elites and economics responsible for global warming, the ground-zero victims of global climate change, or the activist citizens who are leading the battles for climate justice; he makes no connections between a meat-based diet and its environmental consequences. Thus, the film avoids invoking oppressor guilt, though still encouraging action. From a feminist and environmental justice standpoint, Gore's analysis is woefully incomplete. While narratives that inspire environmentally minded action are surely laudable, Gore's limited solutions will not address or rectify all climate injustices.

Science fiction films build on Gore's apocalyptic tenor, but without its rhetorical balance, they actually undermine the credibility of climate scientists. *Waterworld* (1995), *The Day after Tomorrow* (2004), *Artificial Intelligence* (2001), *Elysium* (2013) and *Snowpiercer* (2013) depict white male heroes working to restore life or love against a backdrop of climate change consequences. The Mariner hero of "Waterworld" battles with evil pirates and eventually succeeds in his quest to bring an orphan girl, her female caretaker, and a male hydroelectric power expert, among others, to "Dryland" (Mt. Everest) while the Mariner returns to the sea/frontier; on "Dryland," one assumes, the mundane tasks of sustaining life are unsuited to heroic actions characteristic of the Mariner (aptly cast as Kevin Costner). *The Day After Tomorrow* offers a similar narrative of father-figure rescuing child, as paleoclimatologist Professor Jack Hall tries to save the United States from the effects of climate change and its rapidly returning Ice Age while also trying to save his son Sam, who has taken refuge in the New York Public Library, far north of the line of projected safety from freezing.[7] Both films present climate change consequences as too far-fetched to be credible: the entire planet flooded? The next Ice Age, in a week?

The more complicated films of the five also offer some eerie plausibilities. *A.I.* (for *Artificial Intelligence*) offers an eleven-year-old cyborg boy as a hero whose primary quest is to regain a mother's love. The film raises questions about human identity in a future affected by climate change, and suggests readings of humans as cyborgs, the earth as a rejecting mother, and climate change as the ultimate rejection from the earth/mother—a new twist on mother-blaming. *Elysium* provides a white male hero (Matt Damon) living on the impoverished earth (crowded and overpopulated largely by people of color), whose poverty and romance with a Latina single mother compel him to infiltrate the elites on the Earth-satellite Elysium—led by the evil white woman, ably portrayed by Jodi Foster—to download the entire operating system for the elites' instantaneous health care, and through his suicide transferring that health care to an open democratic access that will save not only those suffering on earth, but immediately heals the dying daughter of his beloved Latina. "Snowpiercer" also depicts class divisions, albeit aboard a Noah's Ark Train speeding on a globe-spanning track encircling an Ice Age planet, conditions produced by the failure of geoengineering in solving the climate crisis. The film's white working-class hero leads a rebellion, breaking through the train's many class divisions and uncovering the hedonism of the first-class elites, the train's exploitation of children, and its violent population reduction methods, all ending in the train's explosion, an avalanche, and the edenic return to a frozen (but soon-to-thaw) world for the security system engineer's teenaged daughter and a preteen boy. The unabashed irony of all these narratives is their race and gender reversal: around the world, it is poor women, rural women, and women of color who are most affected by global climate change effects, and it is women who are working as grassroots heroes to mitigate and adapt to the results of a global environmental crisis created by the world's elites, largely, white men (Women's Environmental Network 2007).

Has ecocriticism been unable to contribute to ongoing conversations about global warming, simply for a lack of worthy literary and cultural artifacts? The few ecocritics who have explored global warming believe so. "Literary writing has not kept pace with the developments in science and public policy pertaining to climate change, peak oil, population pressure, and the food crisis," writes Patrick D. Murphy (2008: 14). "American fiction writers have a rather dim track record on the topic of climate change," Scott Slovic concurs (2008: 109). But I wasn't ready to give up so easily. Still seeking narratives with an awareness of intersectionality, and an approach that could bridge the environmental sciences and the environmental humanities, I turned to ecofeminist theory, environmental justice analyses, critical animal studies, and feminist ecocriticism.

GETTING THE FULL STORY: FROM ENVIRONMENTAL
SCIENCES TO CRITICAL ECOFEMINISM

To say the earth's warming is natural, inevitable, is to dismiss the effect of elite overconsumption and waste. Breathe in the shimmering air exuded by a thousand cars, creeping along crowded freeways. Trace the line of brown diffusing into sky behind the jet airplane, watch smokestacks pumping thick grey clouds into a clotted sky. Breathing, drinking, consuming, my body embodies effects from the rising levels of greenhouse gases, airborne and waterborne pollutants. My body also rides in cars, looks out the window of airplanes, switches on lights and uses computers powered with coal and the energy of choked rivers.

I am part of the problem.

My face flushes with heat.

From the aforementioned intersectional standpoint of feminist environmental justice ecocriticism, climate change can be seen as an environmental justice problem with material consequences for the environmental sciences. In her essay, "From Heroic to Holistic Ethics," ecofeminist and vegan activist Marti Kheel (1993) develops her theory of the truncated narrative, a theory that foregrounds the rhetorical strategy of omission: "Currently, ethics is conceived as a tool for making dramatic decisions at the point at which a crisis has occurred. Little if any thought is given to why the crisis or conflict arose to begin with" (1993: 256). In Western ethics, values are debated on an abstract or theoretical plane, and problems are posed in a static, linear fashion, detached from the contexts in which they are formed: "we are given truncated stories and then asked what we think the ending should be," Kheel explains (1993: 255). Creating "ethics-as-crisis" conveniently creates an identity for the ethical actor as hero, an identity well suited to what Val Plumwood (1993) defines as the Master Model. "Western heroic ethics is designed to treat problems at an advanced stage of their history," Kheel argues, and "run counter to one of the most basic principles in ecology—namely, that everything is interconnected. . . . By uprooting ethical dilemmas from the environment that produced them, heroic ethics sees only random, isolated problems, rather than an entire diseased world view. But until the entire diseased world view is uprooted, we will always face moral crises of the same kind" (1993: 258–59). As an alternative to truncated ethical narratives and heroic ethics, Kheel proposes retrieving "the whole story behind ethical dilemmas," uncovering the interconnections of social and environmental perspectives, policies, economics, and decision-making, including all those affected by the ethical "crisis." With the whole story restored, we can work more effectively for solutions to current ecosocial problems, and prevent others in the future, thereby eliminating the need for heroes—though as Kheel wryly observes, "prevention is simply not a very heroic undertaking" (1993: 258).

Kheel's theory describing truncated narratives helps illuminate the "story" of climate change causes and solutions—stories that surface in the popular media, in science, literature, and culture. Upon initial inquiry, the stories we receive about the causes of climate change are narrated by the environmental sciences, which suggest that climate change is primarily a problem of transportation and energy production; on further inquiry, environmental sciences and environmental politics reveal climate change is a problem exacerbated by processes of industrialized animal-based agriculture, as documented by the United Nations' Food and Agricultural Association report, "Livestock's Long Shadow" (FAO 2006). From environmental politics, we then learn the "subplot" of both these "cover stories"[8] is the deeper problem of first world industrialized nations' over-consumption and waste of global nature and all those associated with nature—indigenous people, the "two-thirds" (or "developing") nations, nonhuman animals and ecosystems. An embedded subplot to the first-world/two-thirds-world narrative is the powerful presence of multinational corporations, whose economic force and global trade agreements have the capacity to overpower democratic decisions made at all levels—city, state, and nation. Finally, at the bottom of these narrative hierarchies lie the inequalities of gender, race and species, which are present at almost every level of society and nation.[9]

Restoring many of the missing components from global warming's truncated narrative, the 27 Bali Principles of Climate Justice (2002)[10] redefine climate change from an environmental justice standpoint, using as a template the original 17 Principles of Environmental Justice (1991) created at the First National People of Color Environmental Summit. They connect the unsustainable consumption and production practices of the North (first world industrialized countries) and the elites of the South (two-thirds world, "developing" countries) with the environmental impacts felt most harshly by those in the South and the impoverished areas of the North. The principles address the categories of gender, indigeneity, age, ability, wealth, and health; they provide mandates for sustainability in energy and food production, democratic decision-making, ecological economics, gender justice, and economic reparations to include support for adaptation and mitigation of climate change impacts on the world's most vulnerable populations. The missing pieces from this statement—the role of industrialized animal agriculture, and the specific climate justice impacts on LGBT people—still need inclusion. With these two additional elements completing the story by correcting its heterosexism and speciesism—in effect, its *humanism*—the intersectional analysis provided by the Bali Principles offers the best articulation for restoring the truncated narrative of climate change.

CLIMATE JUSTICE NARRATIVES

The earth is not a woman, not a single body but billions. I think of heavy white bears, swimming in search of solid ice. Unable to sleep on this hot summer night, I rumble downstairs, get a glass of ice water and hold it to my forehead. I think of migrating birds and butterflies, disoriented by diminished wetlands and early-blooming flowers, and of people sitting on rooftops or suffocating in attics, surrounded by floodwater from broken levees. I think of women who cannot swim, women who wait at home for help that never comes, women who drown attempting to bring elders and children to safety.

We are, all, bodies of knowledge.

The interdependence among human, animal, and environmental health is a knowledge legacy of the feminist, antitoxics, and environmental justice movements of the twentieth century. Beginning with Rachel Carson's (1962) research exposing the links between pesticides and human/animal/environmental health, and augmented by Lois Gibbs' activism exposing childhood cancer clusters in her working-class community atop Hooker Chemical's dump at Love Canal; by the feminist antitoxics movement of the 1980s, the breast cancer and environment activism of the 1990s (i.e., Brady 1991; Clorfene-Casten 1996), and the environmental justice movement of the 1990s and beyond; and by the research linking cancer and environmental toxins as reported in Theo Colborn's *Our Stolen Future* (1996), Sandra Steingraber's *Living Downstream* (1997), and Tyrone Hayes' work on the endocrine-disrupting effects of atrazine (2002)—the scientific and experiential grounds for what material feminist Stacy Alaimo (2008) terms "transcorporeality" are over fifty years strong. Our physical selves are more accurately conceived of as material *flows* rather than discrete and bounded bodies, and thus we must also "reimagine 'climate change' and the fleshy, damp immediacy of our own embodied existences as intimately imbricated," argue Astrida Neimanis and Rachel Loewen Walker: "the weather and the climate are not phenomena 'in' which we live at all—where climate would be some natural backdrop to our separate human dramas—but are rather of us, in us, through us" (2014, 559). This transcorporeality, as Neimanis reiterates, is "embodied, but never essentialised"; it is always articulated "in ways that cannot be dissociated from politics, economics, coloniality and privilege—and my [our] embeddedness therein" (2013, 103).

Given the persistence of feminist, antiracist, and environmental theories confirming our transcorporeality with human and more-than-human natures, and our responsibilities/response-abilities to and with these transcorporeal bodies, I found it hard to explain the absence of women, people of color, and ecoqueers as authors of the cli-fi narratives I had surveyed (though white women do figure more prominently as authors of children's books). As

Michael Zizer and Julie Sze found, mainstream global climate change narratives are characterized by "the elision of specific race- and class-based environmental injustices," and redeploy "traditional U.S. environmental tropes in ways that soft-pedal environmental justice goals" (2007: 387). To date, few ecocritics have asked this question: if global warming narratives in fiction, nonfiction, science fiction, and film alike have been largely the domain of white men, what genre are queers and artists of color using to address global warming? As ecocritic Stephen Siperstein suggests, we must be "willing to expand our vision of the genre" (2014). In climate justice documentaries, short stories, music videos, and popular songs, ecocritics may discover more inclusive and intersectional narratives that restore the truncated narratives of climate change, and offer strategies for mitigating the effects of climate injustices.

Perhaps the first climate justice documentary to reach a global audience, *The Island President* (2011) offers an environmental justice (humanities) counterpart to the "inconvenient truth" of Al Gore's environmental science. This film traces President Mohamed Nasheed of the Maldives as he fights to compel industrialized nations to face up to the impact of their climate-changing emissions on the most low-lying countries in the world. While the people of the Maldives visit neighboring island countries to seek refuge and resettlement possibilities, Nasheed meets with global dignitaries who attend the Copenhagen Climate Summit in 2009 where, at the last moment, he makes a speech that salvages an agreement. Though the Copenhagen summit is widely regarded as a failure, it was the first time that India, China, and the United States had agreed to reducing emissions. This documentary is all the more stirring in light of the fact that in February 2012, President Nasheed was forced to resign under threat of violence, in a coup d'etat perpetrated by security forces loyal to the former dictator, and the dictator's half-brother won the Presidential elections in November 2013. The links between a lack of democracy and climate injustice are clear not just in the Maldives, but throughout the Two-Thirds world.

In the short story "Cayera" by Filipino writer Honorio Bartolomé de Dios (2007), a gay beautician, Bernie, and her friends join a movement against logging operations and the construction of an industrial plant in the agricultural village of San Martin. Although the town's elite know Bernie as a trusted aide, her inclusion in the movement is questioned on the basis of her sexuality. Although she and her friends are mocked by the ecoheterosexual marchers in the rally, Bernie later transforms her beauty parlor into a hiding place for rebels, whom she transforms into women to protect them from the military. "Pageantry and performance thus become sites of resistance," Nina Somera (2009) concludes, arguing that climate change "aggravates longstanding inequalities and peculiar situations that strike one's layers of identities—as a tenant farmer, industrial worker, lesbian mother, landless widow, indigenous woman and so on."

Marvin Gaye's 1971 popular hit, "Mercy, Mercy Me (The Ecology)" was remade in 2006 by the Dirty Dozen Brass Band, giving the song a New Orleans jazz flavor, and an album cover depicting a solitary, naked man pulling a canoe through a flooded urban landscape, unmistakably targeting this song to the events surrounding Hurricane Katrina. Gaye asks, "What about this overcrowded land? How much more abuse from man can she stand?"[11] In the context of global warming catastrophes and climate justice, Gaye's lyrics gain new resonance. Hurricane Katrina was an event that made visible the arrogance of culture's attempt to control nature, along with the indifference of urban planners and engineers to the structures of safety allegedly protecting poor people and people of color—particularly women of color (Seager 2006) and the ways these social hierarchies affect the land, water, and nonhuman animals.

Another exemplary artist, India Arie produced two songs on her "Testimony" Volume 2 album (2009) that address global warming from an intersectional, climate justice perspective. The first song, "Better Way," contrasts the government's response to Hurricane Katrina and the politically motivated war in Iraq against the media manipulation and sheer indifference of elected politicians, particularly then President George W. Bush: "Is it democracy or is it the oil? It's in the news every day, we're a paycheck away, and the President's on the golf course."[12] In the tradition of black spirituals, Arie (like Martin Luther King) positions herself as both Moses figure and feminist lyricist with her refrain, "Let my people go!" Her proposed solution to these and other problems of social, environmental, and climate injustice is a reconception of human identity as interdependent, with inclusivity and care presented as the sole strategy for human survival: "I know there's gotta be a better way, and we gotta find it, we gotta stand together, or we can fall apart."

Another song on Arie's album, "Ghetto" argues for an interdependent self-identity that makes connections across nationality, class, and race. Her work persuasively articulates the problems of global justice to first world listeners (the "you" of the lyrics) by exposing the third world within the first world: "to be hungry in L.A. is just like starving in Bombay. Homeless in Morocco is a shelter in Chicago." Contrasting definitions of the ghetto as "a place of minority, and poverty, and overpopulation," Arie insists that "we live on this earth together" and that from outer space, the whole world can be seen as a ghetto because "it's a small world after all." Arie's lyrics playfully reverse the white supremacist erasure and commodification of difference through the Disneyland "small world" metaphor, and the disembodied "eye in the sky" techno-science metaphor of space exploration, whereby the blue ball of Earth equalizes (and erases) all social hierarchies. Mindful of difference, Arie uses these same dominant metaphors strategically, to draw listeners together.

Another climate justice music video, Kool Keith's ("Dr. Octagon") "Trees Are Dying" (2007) presents an African-American boy age 11 or 12 as its rap narrator, dressed as a schoolteacher in a white shirt, black plaid bow tie, red plaid suspenders, white pants and converse high-tops. The boy poses as teacher and newscaster, standing in front of a blackboard where a map of the United States is chalked over with heat-wave temperatures—186 to 202 degrees Fahrenheit—and a hurricane spiral marking the Gulf of Mexico. The blackboard bursts into flames and his back catches on fire. As he dances and narrates, we see other women and children costumed as trees, stiffly marching through a field of clear-cut stumps, falling backward against headstones, and dancing a ring-around-the rosy circle of death in front of nuclear reactors spewing smoke into the skies. A crew of children dressed as scientists mix smoking fluids and pour them into a planet earth bubble that explodes and rolls away; a frenzy of cars drives across the screen and onto a six-lane freeway, each car with a bull's horns strapped to the hood. A white, high-heeled shoe stamps down a building, introducing a robotic white woman wearing a business suit, acting the role of Godzilla; she attempts unsuccessfully to jam a tree branch into a copier machine, as children in white scientist coats mill around her working at other copying machines, and papers fly in the air. Significantly, the majority of the actor-dancers are black preteen children, the landscape and sky are persistently gray, and refrains such as "apathy kills" and "car-car-carbon dioxide" and "like the elephants, trees are dyin'" repeat in lyrics and in superimposed text. The music video's rich metaphorical twists on popular culture make clear connections among species extinction, deforestation, unsustainable transportation and energy production, corporate-driven colonization of nature, white supremacy, adult supremacy and an apathetic gerontocracy.

Along with an intersectional analysis of the root causes of climate change in social, economic, and global injustices, these hip-hop lyrics and music video offer a sensory reconnection that is unavailable via other media such as literature and cinematic narrative: they make viewers want to dance. In contrast to the immobilizing sense of futility, apathy, or denial often inspired by informational overkill from environmental science and cli-fi literature alike, climate justice musical narratives offer a more inclusive and a more popularly accessible medium, one that energizes its audiences and invites movement toward action and activism alike.

CRITICAL ECOFEMINISM AND CLIMATE CHANGE

In the twenty-first century, contemporary branches of environmental feminist theory—whether "material feminism" (Alaimo & Hekman 2008), "eco-ontological feminism" (Blair 2008), or "eco-ontological social/ist feminist thought" (Bauman 2007)—build on key contributions of ecofeminist and environmental justice perspectives, emphasizing the material reality of our interidentity and the value of embodied knowledge. The essentialism-social constructionism debates of the 1990s misconstrued *ecofeminism* and *essentialism* as synonymous terms, denouncing and discarding both in order to emphasize the shaping forces of culture in organizing human identity and experiences.[13] Twenty years later, feminists are acknowledging that the suspect term was *essentialism*, and now argue for a "feminist politics of the-body-in-place . . . founded in an affirmation of our dependence on the earth" (Mann 2006, 129). Material feminism, write Stacy Alaimo and Susan Hekman, explores "the interaction of culture, history, discourse, technology, biology, and the 'environment,' without privileging any one of these elements" (2008, 7). Similarly, critical ecofeminism envisions human identity as continually becoming-with earth others; it involves "see[ing] creativity and agency in the other-than-human world around us" (Plumwood 2009). Thus, "the potential for us to respond meaningfully to climate change," writes Jennifer Blair, will "depend on a re-conception of subjectivity and a re-conception of the ways in which humans perceive and effect change in the material world" (2008, 319). The problem is that "no matter what information about global warming the media communicates, people seem to need to feel the heat themselves in order to respond to the phenomenon in meaningful, change-driven ways" (Blair 2008, 320). The value of climate justice narratives—in documentary, literature, and music—is their capacity to use art to help people "feel the heat" and activate climate interventions.

Certainly, if everyone must experience the effects of climate change first-hand in order to take meaningful action, those actions will come too late to make a difference. Facing the immediacy of climate change, cli-fi narratives have the potential to present not just a techno-science story but rather to narrate our transcorporeality, explicitly addressing differences of gender, race, nation, economics and ecologies, sexuality and species. Only then will cli-fi readers gain a more complete story of climate injustices, and a more effective road map for activist responses.

NOTES

1. For example, Harriet Beecher Stowe's *Uncle Tom's Cabin* (1852) provoked an outcry against slavery, Upton Sinclair's *The Jungle* (1906) mobilized legislation to regulate the meatpacking industry, Rachel Carson's *Silent Spring* (1962) launched the environmental movement and campaigns to ban the use of DDT, and Edward Abbey's *The Monkey Wrench Gang* (1975) launched EarthFirst! and its famous tactics of

"monkeywrenching." For a contemporary call to use narrative as an urgent and timely strategy for climate justice, see Amitav Ghosh's *The Great Derangement: Climate Change and the Unthinkable* (2016).

2. Thanks to the popularity of cli-fi, the field is proliferating at such a rate that any survey will be inevitably incomplete. I recommend Dan Bloom's "Cli-Fi Central" Facebook page, Andrew Dobson's review of cli-fi narratives at http://www.andrew-dobson.com/eco-apocalypse-novels.html and the Eco-Fiction website at http://eco-fiction.com/ for their rich resources.

3. These feminist environmental justice texts have been discussed from an ecofeminist literary perspective (see Alaimo 1998; McGuire and McGuire 1998; Armbruster 1998) and a feminist ecocritical perspective (Grewe-Volppe 2013; Hogan 2013; Stein 2013).

4. Clyde Freeman Herreid (2005) reports on his use of Crichton's *State of Fear* as a primary text for his honors seminar in "Scientific Inquiry: Case Studies in Science." The fact that Crichton's novel is a work of science fiction, not researched fact, did not preclude it from use in Herreid's University Honors Program course at The State University of New York, Buffalo.

5. A notable exception would be the recent publication of Naomi Oreskes and Eric M. Conway's *The Collapse of Western Civilization* (2014).

6. Facts well known to antiglobalization ecoactivists since the battle against the General GATT in the mid-1990s are now summarized in Naomi Klein's *This Changes Everything: Capitalism vs. The Climate* (2014).

7. There are some enjoyable moments to the film, from a progressive ecocritical standpoint: the students in the library are burning Nietzsche's books to stay warm; there's a president who refuses to listen to global warming science, and a vice president who says action will cost too much; and U.S. citizens are portrayed as climate refugees crossing the border into Mexico by fording the Rio Grande, offering a wry commentary on anti-immigrant sentiments.

8. By "cover story" I mean the story that is on the cover of a rather thick stack of narratives about the causes and solutions to global warming. I do not mean that the "cover story" is covering all the narratives, only that it has gained so much prominence that it functions to obscure other intersectional narratives that are also accurate descriptions of experiential fact.

9. I qualify this claim about the omnipresent inequalities of gender and sexualities with the word "almost" to acknowledge the presence of traditional indigenous societies whose gender role differences are unmarked by differential valuations. Such societies are already quite marginalized in the global economy.

10. The principles were developed by international coalition of groups—which includes CorpWatch, third world Network, Oil Watch, the Indigenous Environmental Network, among others—at the final preparatory negotiations for the Earth Summit in Bali in June 2002.

11. The complete lyrics to "Mercy, Mercy Me" can be found at http://www.metrolyrics.com/mercy-mercy-me-lyrics-marvin-gaye.html

12. Complete lyrics to India Arie's "Better Way" can be found at http://www.lyricsmode.com/lyrics/i/indiaarie/better_way.html

13. There are many branches of ecofeminist theory (Gaard 1998; Sturgeon 1997; Merchant 1995), including branches that are liberal, socialist, anarchist, radical feminist, womanist, Africana, and cultural feminist; the latter is the branch most often charged with essentialism. Vegan and vegetarian ecofeminists have argued that the anti-essentialist backlash against ecofeminism was motivated by a deeper backlash against ecofeminism's prescient posthumanist defense of interspecies justice (Gaard 2011).

Chapter 8

Queering the Climate

Are there ecogenders and ecosexualities that could be consistent with a critical ecofeminist praxis? From years of organizing through the "chain of radical equivalences" among social movement actors, advocated by Ernesto Laclau and Chantal Mouffe (1985) as crucial to the formation of a radically democratic social movement, ecojustice activists and scholars have learned the value of deconstructing the role of the Master Self, and providing a location for even those constructed as dominant (whether via race, gender, class, sexuality, or nationality) to embrace a radically ecological vision and stand with—rather than on top of—the earth's oppressed majorities. For any egalitarian socioeconomic and ecopolitical transformation (such as that advocated by ecofeminism) to be possible, both individual and institutional transformations are necessary: both must shift away from over-valuing exclusively white, male, and heteromasculinized attributes and behaviors, jobs, environments, economic practices, laws and political practices, in order to recognize and enact ecopolitical sustainability and ecological genders. Not surprisingly, ecofeminist theory and praxis continues to be articulated by scholar-activists of diverse genders, races, nations, and sexualities. In thinking about futures that don't center white human heteromasculinity and creating a more inclusive and descriptive ecofeminism—one that provides strategies and locations for cocreatively revisioning, educating, and mobilizing those who reject antiecologically gendered constructions—it is useful to start thinking about ecomasculinities, and ecosexualities.

Toxic, hegemonic masculinity requires an ecofeminist rethinking because humans (from industrial capitalists to ecofeminists and environmentalists) are gendered, sexual beings, and gender is crucial to many peoples' erotic expressions. In my own ecofeminism, gender and eroticism are entangled with my love of this earth. I want words for the butch resonance of rhyolite under my fingers when I am rock-climbing. I want language for the erotic attraction arising between my homecoming presence and feline greetings that approach and recede to rub against doorways and chair legs, eyeing me all the while. I want theory for the desire I inhale from a long-limbed lover who smells like trees. And I know I'm not alone in this ecoerotic assemblage of gender, species, and nature. From Virginia Woolf's transgendered protagonist in *Orlando* (Woolf 1928) to Jeanette Winterson's unnamed and diversely gendered narrator in *Written on the Body* (Winterson 1992), queer feminist writers are envisioning ecomasculinities that exceed sexual biologies, encompassing diverse genders, sexualities and sexual practices.[1]

As Annie Sprinkle writes in *Bi Any Other Name*, "I started out as a regular heterosexual woman. Then I became bisexual. Now I am beyond bisexual— meaning I am sexual with more than just human beings. I literally make love with things like waterfalls, winds, rivers, trees, plants, mud, buildings, sidewalks, invisible things, spirits . . . " (Sprinkle 1991, 103).[2] From Sprinkle's position as a former sex worker and porn star, and now self-described "ecosexual" to Terry Tempest Williams' public position as Mormon heterosexual wife and environmental writer, the expression of ecoerotics and ecogenders suggests a wealth of information for considering ecomasculinities. In literature, Terry Tempest Williams' *Desert Quartet* (1995) describes ecoerotic encounters between a human hiker and the four elements. In the slot canyons of Utah's Cedar Mesa, the narrator's palms "search for a pulse in the rocks," while her body finds "places my hips can barely fit through" until "the silence that lives in these sacred hallways presses against me. I relax. I surrender. I close my eyes. The arousal of my breath rises in me like music, like love, as the possessive muscles between my legs tighten and release. I come to the rock in a moment of stillness, giving and receiving, where there is no partition between my body and the body of earth" (Tempest Williams 1995, 8–10). Hiking along a creek in the Grand Canyon, "only an hour or so past dawn," the narrator decides to take off her "skin of clothes" and leave them on the bank, lying down on her back and floating in water: "Only my face is exposed like an apparition over ripples. Playing with water. Do I dare? My legs open. The rushing water turns my body and touches me with a fast finger that does not tire. I receive without apology. Time. Nothing to rush, only to feel. I feel time in me. It is endless pleasure in the current" (Tempest Williams 1995, 23–24). Are the rock and the water gendered in these ecoerotic encounters? Or does the ecoerotic include and trans* gender? For as the coeditors of *TSQ*'s special issue on "Tranimalities" explain, "a trans* heuristic allows us to better understand the limits of 'the human' as a biopolitical tool for privileging a few so as to de-, in-, nonhumanize the many" (Hayward and Weinstein, 2015, 195).[3]

Perhaps it is (past) time to envision alternative genders and sexualities from an ecofeminist perspective. What would it mean to redefine, or reconceive, an ecological masculinity?

TOWARD AN ECOLOGICAL MASCULINITY

Many ecofeminist philosophers, men's movement writers, animal studies and cultural studies scholars offer diverse yet mutually reinforcing critiques of Euro-Western cultural constructions of masculinity as predicated on themes of maturity-as-separation, with male self-identity and self-esteem based on dominance, conquest, workplace achievement, economic accumulation, elite consumption patterns and behaviors, physical strength, sexual prowess, animal "meat" hunting and/or eating, and competitiveness. These constructions developed in opposition to a complementary and distorted role for women: white hetero-human-femininity (Adams 1990; Buerkle 2009; Cuomo 1992; Davion 1994; Plumwood 1993; Schwalbe 2012). Recent studies of hegemonic masculinity as portrayed in men's lifestyle magazines confirm its pervasive representation via discourses of appearances (strength and size), affects (work ethic and emotional strength), sexualities (homosexual vs. heterosexual), behaviors (violent and assertive), occupations (valuing career over family and housework), and dominations (subordination of women and children) (Ricciardelli, Clow, & White 2010, 64–65). These representations varied far less than researchers expected among straight and gay-oriented men's magazines, and reaffirm the continuing force of hegemonic masculinity across sexualities and nationalities.

A term that was named "word of the year" in 2003 (Danford 2004), *metrosexuality* was articulated via the mass media television show, *Queer Eye for the Straight Guy*, and *The Metrosexual Guide to Style* by Michael Flocker (2003); both productions promoted "gay" advice for heterosexual men, emphasizing "self-presentation, appearance, and grooming" (Ricciardelli, Clow & White 2010, 65). Beneath metrosexuality's "softened" masculinity, scholars found the same hegemonic masculinity, influenced and intensified by consumerism, youth-obsession, and an emphasis on appearance, mandates usually enforced for femininity, and by association (Pharr 1988), for gay males as well: "whenever hegemonic masculinity is challenged, a new hegemonic form emerges," and thus "hegemonic masculinity actually becomes more powerful because of its ability to adapt and to resist change" (Ricciardelli, Clow & White 65).

From bikini waxing to collagen injections and shopping (Frick 2004), metrosexuals were soon called back to hegemonic heteromasculinity through beef consumption. Discussing Burger King's commercial "Manthem" as a textual narrative of gender, C. Wesley Buerkle (Buerkle 2009) reaffirms arguments made by Carol Adams (Adams 1990) that the very act of eating is associated with masculinity, and meat-eating is an act of masculine self-affirmation.[4] Through advertising images and commercials, fast-food franchises such as Burger King and Hardee's portray hamburger consumption as enacting men's symbolic return to their supposed essence: personal and relational independence, nonfemininity, and virile heterosexuality. Continuing research reaffirms and uncovers further evidence to support Marti Kheel's critique of hegemonic masculinity's antiecological foundations, articulated in the ways it "idealizes transcending the [female-imaged] biological realm, as represented by other-than-human animals and affiliative ties" and "subordinate[s] empathy and care for individual beings to a larger cognitive perspective or 'whole'" (Kheel 2008, 3). The unstated fact that the more-than-likely "spent" dairy cows and other juvenile cattle slaughtered for fast-food beef hamburgers served at Burger King and Hardee's contribute exponentially to the accelerated rate of global warming (Steinfeld et al. 2006; Springmann 2016) underscores the antiecological impact of beef-eating hegemonic masculinity.

As the many mass killings and terrorist attacks have made clear, hegemonic masculinity is lethal—for children (Columbine High School in 1999, Sandy Hook Elementary School in 2012), for women (the Montreal Massacre of women engineering students in 1989), for queers and Latinx (Pulse Nightclub in 2016), for people who dine out or travel or celebrate holidays (Paris in 2015, Brussels airport in 2016, Nice for Bastille Day in 2016), and finally, for the men themselves, as terrorists either commit suicide or are shot by police.[5] In the United States, the majority of terrorist attacks are carried out by white men, not by Muslims or Arabs: one study documents that from 1982 to 2016, forty-four of the mass killers were white males; one was a woman (Follman, Aronsen & Pan 2016). Except in two cases, the killings were carried out by a lone shooter, a man acting alone in what the Black community of the Planned Parenthood Action Fund (and others) call "toxic masculinity" (Berry 2016; Ochoa 2016). Research suggests that "mass shooters experience masculinity threats from their peers and, sometimes, simply from an inability to live up to societal expectations associated with masculinity (like holding down a steady job, being able to obtain sexual access to women's bodies, etc.) . . . . Research does not suggest that men are somehow inherently more violent than women [but rather] that men are likely to turn to violence when they perceive themselves to be otherwise unable to stake a claim to a masculine gender identity" (Bridges and Tober 2015). This public display of masculinity-as-violence is often companioned by the private display of domestic violence, though it's only the public displays that get counted as terrorism. The shooter at the Pulse nightclub in Orlando, Omar Mateen, reportedly beat his first wife and at one point held her hostage— but he was never held accountable for his actions, as under pressure from her family, the woman divorced him after only four months of marriage. According to the Federal Bureau of Investigation reports, between January 2009 and 2014, 57% of mass shootings involved a perpetrator killing a partner or other family member: "in other words, men killing women intimates and their children and relatives are the country's prototypical mass shooters" (Chemaly 2016). While public mass shootings are terrifying because the victims are unsuspecting and unrelated to the killers, in domestic violence, the perpetrators are well known. But in both contexts, extreme violence (mass killing) is a pathology that reaffirms masculinity-as-dominance, defined through its contrast with a hated, feminized vulnerability and its counterpart—homophobia (Pharr 1988). That these violent behaviors of hegemonic masculinity are antiecological (i.e., anti-life) clarifies the urgency of exposing this gender-hegemony and exploring alternate, ecological forms of masculinity instead.

But masculinity has not always been defined in opposition to ecology. Although Lynn White's (1967) critique of Christianity's anthropocentric dominion over nature is probably the first and best-known, feminist and ecofeminist theologians such as Rosemary Radford Ruether (1983, 1992), Carol Christ (1997, 1979), Charlene Spretnak (1982), and Elizabeth Dodson Gray (1979) advanced beyond White's critique, offering significant critiques of monotheistic, patriarchal religions that worship a sky god and remove spirituality and the sacred from the earth, placing hell beneath our feet and heaven in the sky, deifying men, and valuing men's associated attributes over the values, attributes, and bodies of women, children, nonhuman animals, and the rest of nature. But prior to patriarchal, monotheistic religions, history and archeology show a different value was placed on women, nature, fertility, and the cycles of the earth. Following feminist theologians, men's movement scholars interested in mythology and archetypes make an important distinction between "sky god archetypes [who are] often warlike, either youthful invincible heroes, or older dominant males who rule in the name of an all-powerful [and often wrathful] sky god" and who serve to define masculinity as "a journey of ascension," and contrast them with earth gods: "archetypal images that relate masculinity to the earth" and offer "a different journey, one of descent, a 'going down' into, initially for many men, grief" (Finn 1998). In Arizona and New Mexico, the earth god is Kokopelli, the hump-backed flute player, a 3,000-year-old Hopi symbol of fertility, replenishment, music, dance, and mischief. In Europe, it is the Green Man, pictured as a male head disgorging vegetation from his mouth, ears, eyes; often associated with serpents or dragons, the Great Goddess, and the sacred tree, the Green Man dates back to Celtic art before Roman conquest, and to the work of Roman sculptors in the first century C.E., and includes manifestations in figures such as Osiris, Dionysus, Cernunnos, and Okeanus (Anderson 1990).

As many feminist spirituality groups have discovered, most people can't jump backward in history, and attempts to re-enter and revive ancient traditions can seem not only ill-fitting, but also fail to provide maps and solutions for contemporary ecosocial problems. Their value, however, lies in the fact that their presence proves there have been ecological, life-giving, and nurturing attributes associated with masculinity; thus, alternatives to hegemonic, antiecological masculinism may again be possible. Reconstructing ecological masculinities in 2016 and beyond, however, will require influence and insight from the last century of ecojustice movements, philosophies, and activisms.

Certainly, there have been significant silences that need to be addressed. As Mark Allister explains in *Eco-Man*, "gender studies in ecocriticism have been dominated by attention to feminism, [and] men's studies has been blind in seeing nature," citing the most important anthology in the field, Michael Kimmel and Michael Messner's *Men's Lives,* for supporting evidence (Allister 2004, 8–9). Yet despite its provocative subtitle, and its expressed intention "to serve as a companion to ecofeminism," *Eco-Man: New Perspectives on Masculinity and Nature* offers "no consistent underpinning" for its contents, and "no general deconstruction . . . of masculinity" (Allister 2004, 8). Similarly, in the premier volume on *Queer Ecologies* (Mortimer-Sandilands and Erickson 2010), while ample focus is given to human queer identities and other species' queer sexual practices, no attention is paid to the practices and organizations inspired by vegan lesbians, the presence and meaning of numerous websites and listserves for queer vegetarians, or the argument that vegan sexuality challenges heteronormative masculinity (Potts and Parry 2010); moreover, discussions of gender are relegated to a footnote summarizing Judith Butler's (1997) description of masculinity and femininity as "precarious achievements that are socially and psychically produced, in the context of a prohibition against homosexuality, through the *compulsory loss* of homosexual attachments," a loss that is "essentially melancholy in character" (Mortimer-Sandilands and Erickson 2010, 356, n. 10). In sum, neither ecocriticism, nor men's studies, nor queer ecologies, nor (to date) ecofeminism has offered a theoretically sophisticated foray into the potentials for ecomasculinities (not to mention transecologies).[6]

Perhaps this omission is rooted in second-wave feminism's rejection of gender roles as universally oppressive. Radical texts of second-wave feminism such as June Singer's *Androgyny* (1977), which explores diverse religious and philosophical traditions from Plato's *Symposium* to the Book of Genesis, and from Jewish mysticism in the Kabbalah to Hindu practices of Tantra, concludes that both masculine and feminine traits are fragmented and distorted parts of a whole and healthy psyche (meaning both soul and mind), and our job as self-actualizing humans involves "transcending" gender and "simply flowing between the opposites" (Singer 1977, 332). Yet such thousand-year-old constructions of gendered identity perpetuate the notion of dualized and polarized gender characteristics, advancing an essentialism that ecofeminists later rejected as limiting to theory-building and inclusivity (Davion 1994; Cuomo 1992). Instead of perpetuating the heterosexually distorted binary gender roles of masculine and feminine through an ideal of androgyny, or pretending that gender can be erased by eschewing all gendered cultural practices (from shaving and make-up to competition and weight-lifting), feminist ecomasculinity theorists need to reconceive gender—because we can't dismiss it. As a primary portal to the erotic, gender is more engaging when multiply expressed and freely crafted into diverse expressions. Moreover, exploring gender quickly leads to exploring sexualities, and opens the possibilities for not just ecomasculinities but ecogenders, ecosexualities, and the ecoerotic.

As Jack Halberstam's groundbreaking volume on *Female Masculinity* (Halberstam 1998) describes, there is a long history of women-born-women with variously expressed masculine gender identities, from nineteenth-century tribades and female husbands to twentieth-century inverts and butches, transgender butches and drag kings. But the conjunction of *ecological feminist* politics and practices with these female masculinities has not been fully theorized, and the conjunctions range widely: while some articulations of female masculinity—most notably, some drag kings and transmen—have perpetuated oppressive manifestations of masculinity via sexism, exercising male privilege, and objectifying women, other articulations of transfeminist masculinities, exemplified in the annual Cascadia Trans and Womyn's Action Camp, engage in everything from "ecosexual hikes" to "climb line rigging," "coalitions against coal," "nonviolent action training," "racism through an intersectional framework" and "self-care for cyclists" (Trans & Womyn's Action Camp 2012). Halberstam's research addresses the embodiment of female masculinities as performance and identity, providing groundwork for further attention to intellectual forms of female masculinities (i.e., theory-building, interrupting/contesting the corporate media), to the ecopolitical relation between butch identities and veganism, for example, or to climate justice and the material realities of economically marginalized women, people of color, queers, and nonhuman animals. As even Halberstam writes in their concluding chapter, "I do not believe that we are moving steadily toward a genderless society or even that this is a utopia to be desired" (Halberstam 1998, 272). Instead, theorizing the ecological articulations of a diversity of genders and sexualities may be a more strategic way to explore material dimensions of animal and ecological health; although lesbian femme and heterofeminine genders also have ecological intersections, I am more drawn to exploring ecomasculinity because masculine gender identity has been constructed as so very *anti*ecological, and thus its interrogation and transformation seem especially crucial. Moreover, the tools for this exploration are close at hand.

From ecofeminist theory, "boundary conditions" for ecomasculinities can be adapted to offer preliminary groundwork. For example, Karen Warren's eight boundary conditions of a feminist ethic (Warren 1990), as applied to ecomasculinity, might read (1) not promoting any of the "isms" of social domination; (2) locating ethics contextually; (3) centralizing the diversity of women's voices; and (4) reconceiving ethical theory as theory-in-process which changes over time. Like a feminist ethic which is contextualist, structurally pluralistic, and in-process, an ecomasculinity would also strive to be (5) responsive to the experiences and perspectives of oppressed persons of all genders, races, nations and sexualities; (6) it would not attempt to provide an objective viewpoint, knowing that centralizing the oppressed provides a better bias. As with feminist ethics, an ecomasculinity would (7) provide a central place for values typically misrepresented in traditional ethics (care, love, friendship, erotics), and most significantly, (8) reconceive what it means to be human, "since it rejects as either meaningless or currently untenable *any gender-free or gender-neutral description of humans, ethics, and ethical decision making*" (Warren 1990, 141, italics mine). By rejecting abstract individualism, feminist ecomasculinities would recognize that all human identities and moral conduct are best understood "in terms of networks or webs of historical and concrete relationships" (Warren 1990, 141). Building on Warren's theory, an ecological masculinity would have to be explored through cross-cultural and multicultural perspectives to protect against privileging any specific race, region, or ethnicity. Patriarchy has shaped most contemporary industrial-capitalist cultures, so ecomasculinities would need to recognize and resist the identity-shaping economic structures of industrial capitalism, its inherent rewards based on hierarchies of race/class/gender/age/ability/species/sex/sexuality, and its implicit demands for ceaseless work, production, competition, and achievement. With ecofeminist values at heart, ecomasculinities would develop beyond merely rejecting the bifurcation of heterogendered traits, values, and behaviors: ecomasculinity/ies would enact a diversity of ecological behaviors that celebrate and sustain biodiversity and ecological justice, interspecies community, ecoeroticisms, ecological economics, playfulness, and direct action resistance to corporate capitalist ecodevastations. Already, developments are underway.

To date, Paul Pulé (2007, 2009) has been foremost in developing an "ecological masculinism" that replaces an "ethic of daring" (based on dominant male values such as rationality, reductionism, power and control, confidence, conceit, selfishness, competitiveness, virility) with an "ethic of caring" for self, society, and environment (with associated values of love, friendship, trust, compassion, consideration, reciprocity, and cooperation with human and more-than-human life). Optimistically, Pulé identifies eight key conceptual frameworks across the political spectrum, along with seven "liberatory ideals in sympathy with Leftist politics" that he believes support "a shift away from hegemonic masculinities and towards a long-term ecological sustainability"; he proposes an ecomasculinism that "may crucially contribute to this shift" (Pulé 2007).[7] While Pulé's work offers a foray into this discussion, he omits Plumwood (1993), Warren (1994, 1997, 2000), Salleh (1984, 1997), and many other ecofeminist critiques of several of his listed key conceptual frameworks and liberatory ideals—critiques that prove many of these conceptual frameworks to be inherently unsuited to even a *feminist* revisioning of ecologically oriented gender. Moreover, apart from a footnote, Pulé does not consider the strong influences of race, class, sexuality, and culture in constructing masculinities.

In advancing a truly ecological and feminist masculinity, the heterosexism implicit in hegemonic constructions of masculinity would need to be resisted (cf. Hultman 2013; Anshelm and Hultman 2014) drawing on insights and questions from the new queer ecologies (Gaard 1997; Mortimer-Sandilands & Erickson 2010). What would an eco-trans-masculinity look like? Are all Lesbian Rangers ecobutches, or are there ecofemmes flashing lesbian masculinities in the Parks Service as well?[8] Could we imagine ecofags, radical faeries who dance and flirt and organize for ecosexual justice?

FEMINIST ECOGENDERS AND ECOSEXUALITIES

Indeed we can. In the 1960s and 1970s, gay liberation activists seeking ways to articulate the intersections of gay sexuality, spirituality, ecoanarchist politics, and genderfuck created the Radical Faeries. Describing themselves as "a network of faggot farmers, workers, artists, drag queens, political activists, witches, magickians, rural and urban dwellers who see gays and lesbians as a distinct and separate people, with our own culture, ways of being/becoming, and spirituality," Radical Faeries believe in "the sacredness of nature and the earth [and] honor the interconnectedness of spirit, sex, politics and culture" (Cain & Rose). They include legendary queer visionaries such as Harry Hay, Will Roscoe, and Mitch Walker, with their history and vision articulated through Arthur Evans' *Witchcraft and the Gay Counterculture* (Evans 1978). Within their regionally placed communities and annual gatherings, Radical Faeries celebrated an earth-based spirituality that honored sexuality and began the applied work of articulating contemporary, nonhegemonic ecomasculinities. Describing themselves as "not-men," "sissies" and "faeries," the Radical Faeries "manifesto" offered only a short statement about feminism: "As faeries we are very interested in what our sisters have to say. The feminist movement is a beautiful expansion of consciousness. As faeries we enjoy participating in its growth" (Cain & Rose). Unfortunately, the faeries continue to describe the earth as female, a gendering that ecofeminists have shown tends to perpetuate Eurocentric gender stereotypes (i.e., earth as nurturing mother who will clean up men's toxic wastes, as a bad and unruly broad who brings hurricanes and other "bad" weather, or a virgin to be ravished/colonized) and does not improve real material conditions for women or nature (Gaard 1993). Nonetheless, the Radical Faerie movement launched interrogations of queer ecomasculinities that have been advanced over the past four decades.

In "Wigstock" (1995), a documentary covering the annual drag festival in New York City, emcee Lady Bunny says, "I think Mother Nature must be a Drag Queen," articulating the nexus of ecomasculinity and genderfuck that drag queens are well positioned to provide. Consider the "radioactive" drag queen Nuclia Waste (Krupar 2012), whose flamboyant performances draw attention to the cleanup efforts at former plutonium production facility Rocky Flats, Colorado, that have converted this location into a wildlife refuge. As a triple-breasted and sparkly bearded drag queen with glowing green hair, Nuclia Waste makes visible "the porosity of body and environment and the ways humans and nonhumans have been irrevocably altered by nuclear projects" (Krupar 2012, 315). Her digital performances reintegrate toxic waste, mutant sexualities, and popular culture, "queering the nuclear family" and encouraging viewers to consider "the entire US as a nuclear landscape" and the pervasive "presence of nuclear waste in everyday life" (Krupar 2012, 316). Mixing stereotypically male and female signifiers, Nuclia's trans* performances enact an irreverent critique of nature/culture and waste/human binarisms, insisting on the impossibility of purity and queering humans as "boundary-creatures, neither fully natural nor fully civilized" (Krupar 2012, 317). Her work could also be used to address the real material conditions for "wildlife"—that is, the more-than-human animals reintroduced to clean up appearances at this nuclear waste site. In sum, Nuclia's performances offer an embodied ecological politics that is crucial to an ecofeminist reconsideration of ecogenders and ecosexualities—and, interspecies ecologies need to be central in such reconsiderations.

Introducing the term "ecogender," Banerjee and Bell (2007) argue that "women and men have been interacting with the environment for ages, *qua women and men*, without consciously attempting to do so" (3). Although their research accepts sexual and gender dualisms, they offer an environmental social science critique of ecofeminism that is helpful to this project of constructing ecomasculinities: "Merchant's [1980] view of precapitalist society passes easily over the brutality of feudal hierarchies," they observe, and "Plumwood [1993] does not identify the logic of domination outside of the West" even though patterns of dominating women, non-dominant men, children, more-than-human animals and nature can readily be found in non-Western societies. Moreover, "Mellor's [1992, 1997] vision of women as environmental mediators homogenizes women's experience and unnecessarily excludes men as potential mediators," and "Salleh [1984, 1997] does not confront the question of the commodification of men and male labor" (Bannerjee and Bell 2007, 7–8). Far from articulating the antifeminist complaint that "men are oppressed too!", Banerjee and Bell remind us that the elevation of a few elite men has been advanced at the expense of other less-dominant men, women, children, animals, and the environment. As Warren's boundary conditions suggest, liberatory theories that exclude or overlook the oppression of any subordinated group cannot hope to provide a holistic description of the logic and functioning of oppressive systems, or propose effective strategies for their transformation. Based on the understanding that "gender itself is a relational construction, and that therefore women's and men's embodied environmental experience cannot be understood in isolation" but must be historically and culturally situated, Banerjee and Bell propose an ecogender studies to explore "the dialogic character of the relationality of gender, society, and environment" which will uncover "the patterns of oppression that constrain these interactions" (Bannerjee and Bell 2007, 14). Although their study omits consideration of sexualities and species relations, their articulation of ecogender as an encompassing approach to bringing ecofeminist theory into the environmental social sciences directly addresses Kheel's critique of hegemonic masculinity.

Putting these diverse approaches together raises questions about the ecological implications and entanglements of gender and sexuality alike. Approaching ecogender from the perspective of bisexuality, Serena Anderlini-D'Onofrio argues that our "current erotophobic cultural climate" can be disrupted by bisexual practices that function "as a portal to a world without the homo-hetero divide," unleashing an erotophilia whose "transformative force" can power more loving and ecologically effective responses to climate instability and a variety of human health crises (Anderlini-D'Onofrio 2011, 179, 186, et passim). To move beyond the bedroom and into the sociopolitical realm, this erotophilia needs to be linked with a critical ecofeminist political approach in order to bring its "transformative force" forward, from the erotic to the ecoerotic.[9]

Clearly, the humanist orientation of most culturally constructed masculinities must be interrogated. Disentangling biological sex, gender expression, gender role, sexual orientation, and sexual practices, as queer studies scholars like Anderlini-D'Onofrio and others suggest, can we describe (not define) diverse ecosexualities that play fast and loose with gender while actively working for environmental, interspecies, and climate justice? How might a queer, interspecies consideration of gender guide our revisioning of human ecomasculinities and ecosexualities?

CONTEMPORARY ECOMASCULINITIES

For examples of ecomasculinities and the ecoerotic, we can look to the music and lyrics of "Nature Boy" eden ahbez, ecocritic Jim Tarter, and Sami-American artist Kurt Seaberg.[10] In 1947, eden ahbez (who always spelled his name in lower case) approached Nat King Cole's manager in Los Angeles, and handed him the music and lyrics for "Nature Boy," a song that quickly became famous. A disciple of Paramahansa Yogananda's silent meditation practices, ahbez lived a life of economic and ecological simplicity, wearing burlap pants and sandals, sleeping outdoors underneath the Hollywood sign, and eating a vegetarian diet. He later lived in community with other like-minded yogis in Laurel Canyon, and collaborated with jazz musician Herb Jeffries on his "Nature Boy Suite." Predating the hippie movement of the 1960s, ahbez performed bongo, flute, and poetry gigs at beat coffeehouses in the Los Angeles area. In 1960—the year I was born—ahbez recorded his only solo LP, *Eden's Island,* for Del-Fi Records. Growing up in a Los Angeles suburb in the 1960s, I listened to eden ahbez' album "Eden's Island" and played over and over again "The Wanderer" for its ecological economics, ecospirituality, and trans* species identity, in lyrics reminiscent of Walt Whitman's poetry:

> *To live in an old shack by the sea*
> *(And breathe the sweet salt air)*
> *To live with the dawn and the dusk*
> *The new moon and the full moon*
> *The tides and the wind and the rain . . .*
> *And lose all sense of time*
> *And be free . . .*
> *And in the evening*
> *(When the sky is on fire)*
> *Heaven and earth become my great open cathedral*
> *Where all men are brothers [sic]*
> *Where all things are bound by law*
> *And crowned with love*
> *Poor, alone and happy*
> *I make a fire on the beach*
> *And as darkness covers the face off the deep*
> *Lie down in the wild grass*
> *And dream the dream that the dreamers dream*
> *I am the wind, the sea, the evening star,*
> *I am everyone, anyone, no one.*

It wasn't until adulthood that I learned ahbez had been living two miles from my childhood home, where his ecomasculinity, ecospirituality, and interspecies ethics preceded and made space for my own ecofeminist ethics.

Another vibrant example of feminist ecomasculinity can be seen in the life of ecocritic Jim Tarter. Writing in *The Environmental Justice Reader*, Tarter (2002) describes his battle with Hodgkin's Disease, a cancer of the lymphatic system, and his sister's battle with ovarian cancer. Quitting his career-track job, Tarter moved in with his sister Karen and became her primary, live-in caretaker for the last six months of her life. Together, they read Sandra Steingraber's *Living Downstream: An Ecologist Looks at Cancer and the Environment* (1997), and pieced together their family's battle with cancer in conjunction with the toxic environment of their early childhood years along the Saginaw River in Michigan, with General Motors and Dow Chemical and cement factories nearby. Through his caregiving of Karen, and their readings of Steingraber, Tarter realizes that cancer is an environmental justice issue for the ways it affects women's bodies, with the most dangerous carcinogens stored in body fats, and the cancers attacking women's reproductive organs (breast, uterus, ovaries). After Karen's death, Tarter continued his teaching and ecocritical scholarship in Idaho with this new focus, committing his teaching to educating indigenous students and bringing feminist perspectives into his work.

Lithographer, carpenter, gardener, playful actor and ecoactivist writer, Kurt Seaberg makes his home by the Mississippi River, in a duplex he shares with poet Louis Alemayhu. Seaberg has replaced his entire back yard and drive-way with a sustainable garden, where he grows much of his own food, and has planted native grasses around the front and sides of the home. Songbirds, hummingbirds, and wasps receive equal welcome in his garden with nests and feeders tucked among lattices and eaves, while mice find forage in his compost bins. As a low-tech response to climate change, Seaberg uses his bicycle for transportation, and participates in local activist groups such as Friends of the Mississippi River, MN350.Org, Tar Sands Action, and supports the Indigenous Environmental Network. A former men's group participant and Green Party supporter, Seaberg brings a vision of social and ecological justice into his work, his creative and performing artistry, and his strong community ties. His artist's statement describes this work:

> One of the tasks of the artist, I feel, is to remind us where our strength and power lies—in beauty, community and a sense of place. Nature has always been a theme and source of inspiration in my work, in particular the spiritual qualities that I find there. My hope is that my art will evoke the same feelings that arise in me when I contemplate the mystery of being alive in a living world: humility, gratitude and a sense of wonder before what I believe is truly sacred.

As Schwalbe (2012) reminds us, reconstructing hegemonic masculinity requires crucial actions linking the individual and the institutional: as examples, he suggests nurturing new minds in children, minds not oriented to seeking satisfaction in status, power, and the domination of others, nor in submission or blind obedience; and working to "end the exploitative economic and political arrangements that are sustained by a continuing supply of expendable men" (Schwalbe 2012, 42). Without the need to dominate and control others—and with the creation and ongoing presence of cooperative economic and democratic enterprises—there will be "little need for the kind of manhood that has evolved under capitalism" (Schwalbe 2012, 44). Writing in *Eco-Man*, Patrick D. Murphy suggests another key feature of ecomasculinity: noting that "men are credited with creating but are not expected to nurture what they create," he laments that "nurturing remains a concept rarely applied to men and an area of male practice inadequately studied, discussed, and promoted" (Murphy 2004, 196–97). Murphy's essay explores some of the ways that fathers can nurture children while learning "to relinquish the illusion of control" and engaging with the fathers' own emotions, in dialogue with their children (Murphy 2004, 208). Nurturing ecological sustainability, nurturing human and more-than-human companions, nurturing an ecophilic ecoerotic, and nurturing interspecies, ecological justice: these are some of the projects of a critical feminist ecomasculinity.

TOWARD ECOEROTIC JUSTICE: ECOMASCULINITIES AND ECOSEXUALITIES

Beginning with Cate Sandilands' "Lavender's Green? Some thoughts on queer(y)ing environmental politics" (1994), ecofeminists have explored the intersections of queer theory, feminism, postcolonial, and environmental thought in the United States, discovering that erotophobia and hegemonic heterosexuality are not only part of dominant Western ideas of nature, but also interstructured with environmental degradation (Gaard 1997). Colonialism, white heteromale supremacy, hegemonic heterosexuality, and the linked devaluation of the erotic with those associated with/seen as "nature"—indigenous people, women, nonhumans, queers—intersect to authorize and advance the domination of nature for the benefit of the few. From these insights, it's clear that the intersections of environmental politics, gender, and erotophobia merit investigation.

As Sandilands' (2001) has noted, the *naturalization of heterosexuality* is interlinked with the *heterosexualization of nature,* and together these influence Western culture's erotophobia. From a queer ecofeminist perspective, heteronormativity's "central narrative of sex-death-redemption" is "threatened with the revelation of other erotic possibilities as potentially natural" and in fact "richer and more genuinely affirmative" (Sandilands 2001, 182).

Perhaps anticipating Stacy Alaimo's (2010) concept of transcorporeality, Sandilands suggests an environmental erotogenic practice that involves "a desiring re-mapping of body and self in passionate contingency with an Other," creating "a relational understanding of self" (2001, 186). Her argument affirms that the ecological crisis is, in part, a problem of the "narrowing, regulation, erasure, ordering, atomization and homogenization" of desire (2001, 188).

In the new millennium, more queers have insisted on bringing their sexual and ecological passions together, and ecoqueer organizations have flourished: the gay and lesbian Sierra Club chapters in California, Colorado, and Washington; the Washington-based Out4Sustainability, the U.S.-based Queer Farmer Film Project, San-Francisco-based Rainbow Chard Alliance, Toronto's Ecoqueers, Minnesota-based Outwoods, and more (Sbicca 2012; Gosine 2001). Flash forward to September 2014 and the People's Climate March in New York City, where 17 people gathered at the "Queers for the Climate" workshop (the only queer workshop out of 119) to brainstorm ways to bring a queer social movement repertoire to work for climate justice.[11] Their answers included many characteristics and skills queers offer: a love of beauty, an appreciation for nonviolence, the understanding of intersectional oppressions (exacerbated in times of crisis), the urban/rural schism in queer communities, marriage equality victories and challenges, AIDS mobilization skills, and the joys of queer performativity, pageantry, drag, and polymorphous perversity. Performance artists Beth Stephens and Annie Sprinkle address all these foundational elements in queer ecofeminist and climate activisms through their ecosexual documentary challenging coal mining and mountaintop removal, "Goodbye Gauley Mountain" (2013).

Queering Relational Ontologies. Visualize this succession of images: a dirt-blackened hill covered with digging trucks. Bees on flowers. Butterflies on milkweed. Mountains exploding. "I used to think the mountains would be there forever, and I would retire in their embrace," Beth Stephens tells viewers.

Feminist relational ontology suggests we are born and come into being through relationships, and these relationships are not only human-to-human but also human to more-than-human, including relations with other animals, plants, waterbodies, rocks, soils, and seasons. In the first section of "Goodbye Gauley Mountain," Beth Stephens reconnects with the places where she grew up, the mountains and streams, forests and towns, friends and family that together form her ecocommunity of origins. She recalls her family's roots in West Virginia's mountains and in coal mining, establishing that she has both a right and a responsibility to stand with this community, to bring her whole self and all her relations to the struggle for environmental justice. "The places where people are born are genetically imprinted on people, in their psyche and in their hearts," Stephens explains. "So all the people who had connections here [in Lindytown, WV, a small mining community which Massey energy bought up and forced the residents to leave] have been absolutely erased." To heal this rupture and to revive by eroticizing people's material connections with earthothers, Stephens and Sprinkle enact ecosexual polyamorous and posthumanist weddings.

Queering Eros, Thanatos, and the EcoCycles of life. "We used to be lesbians, but now we are ecosexuals," says Stephens. A whole new movement of ecoqueers is emerging from the back-to-the-land lesbian communities of the 1970s, and the gay men cruising one another in city and national parks, on beaches and river gorges (Mortimer-Sandilands and Erickson 2010).

Annie Sprinkle defines ecosexuality as "a way to create a more connected relationship" with earth. "We like to have skygasms," says Sprinkle. "Beth and I have intercourse with the air that we breathe." Their Ecosex Manifesto[12] elaborates:

> We shamelessly hug trees, massage the earth with our feet, and talk erotically to plants. We are skinny dippers, sun worshippers, and star gazers. We caress rocks, are pleasured by waterfalls, and admire Earth's curves often. We make love with the Earth through our senses. We celebrate our E-spots. We are very dirty. . . . We are everywhere. We are polymorphous and pollen-amorous.

The Ecosex pledge reconnects eros and thanatos in a ecocycle of sensual embrace: "I promise to love, honor, and cherish the earth until death brings us closer together forever."

Queering Environmental Justice. Most obviously, mountaintop removal (MTR) is about intersections of race, class, and rurality. Stephens takes care to review the history of black miners who died of silicosis in the 1930s, while later the "Hawks' Nest Tunnel" became the Hawks Nest golf course, with the clubhouse built on the site of shanties which housed hundreds of black workers. Mining companies have long manipulated the poor hillbillies into working for the mine as the only job in town. The film shows corporate mining rallies, where CEO spokesmen preach propaganda to the workers, but later sell the company after an explosion kills twenty-nine workers, while CEOs go on to retire with an $86 million bonus—and start a new company. Appalachians have often been looked down on, considered illiterate, uneducated, unimportant, even disposable. "So who cares if a few hillbillies have to be moved out of their homes or end up with cancer, so that [urbanites] can have cheap electricity?" asks one frustrated activist. This is the urban/rural schism dividing queer communities and environmental communities alike. "I don't think people really care about Appalachia," Stephens reflects. "I believe that coal mining has become a protracted form of genocide," a form of slow violence (Nixon 2011) that is rural environmental classism, wiping out the culture and ecocommunities of West Virginia, where a monoeconomy keeps people in thrall to one industry. Bringing queer performance artists to West Virginia's embattled mining communities, Stephens and Sprinkle's ecosexual weddings bridge the urban/rural, queer/straight schisms by affirming a shared and longstanding love of the mountains, and celebrating that love in drag and polyamorous commitment.

Queering Beauty. What produces ugliness, if not the renaming of ecologically diverse forests and streams as "overburden," or the employment of "disposable" black men who will breathe in silica dust while digging a tunnel to mine and transport coal? Such naming authorizes the scraping away of human and more-than-human lives to extract the coal-black dirt that can be blasted and hauled, creating ill health for all. As the filmmakers inform us, communities near MTR have a 50% increase of cancer, and are 42% more likely to have children born with birth defects. In queer strategic response, "Goodbye Gauley Mountain" recuperates an aesthetic of diversity-in-ecocommunity where interconnections appear in reciprocal flows. Hillbillies and urban queers come together to love, honor, and cherish all vibrant matters of transspecies ecocommunity.

Queering Ecometaphors. What metaphors are sufficiently slippery to queer human relations with more-than-human natures? Ecocriticism is rooted in *The Lay of the Land* (1975), Annette Kolodny's feminist literary critique of heterosexualized metaphors: early explorers, colonists, missionaries, and scholars described the land as "virgin" and its exploitation as "rape." The "mother nature" metaphor in Euro-Western culture was simultaneously revered and vilified, blamed and discredited as human mothers are in that culture, taken for granted and even hated. Whether as Mother, Virgin, or "blue marble" seen from space, each metaphor defines Earth's *telos* in humanist terms of instrumental value. Informed by Kolodny's work, later feminist writings about nature drew on the "sisterhood" metaphor of 1960s feminism, producing texts such as *Sisters of the Earth* (Anderson 1991; 2003) and *Sister Species* (Kemmerer 2011).

In their work advancing Ecosexuality, Stephens and Sprinkle seize the opportunity to bring forward tools of queer feminist ecoactivism to the mountains of West Virginia and beyond. "We shift the metaphor from earth as mother to earth as lover," says Annie Sprinkle, "to entice people to have more love of the planet."

CONCLUSION

Ecofeminism illuminates the ways that white capitalist heteromasculinity is fundamentally antiecological. If environmental ethicists and activists want to make more conscious choices about the gendered lenses through which we view environmental activisms, we'll need to envision diverse expressions of ecogenders—not just ecomasculinities but also ecofemme and ecotrans identities—as well as ecosexualities. As with our cultures, our physical, erotic animal bodies are locations of knowledge to be explored.

NOTES

1. For an ecofeminist ecocritical discussion of such gender-bending literatures, see Justyna Kostkowska, *Ecocriticism and Women Writers* (2013).

2. I excerpted this portion of Sprinkle's sentence because she completes her list by including "beings from other planets, the earth, and yes, even animals" raising the significant question of consent, and how one would determine consent from another species. Consent is a nonnegotiable premise for all radical sexualities and sexual behaviors, from feminist sexualities to queer, polyamorous, and sado-masochistic sexual behaviors. Both Annie Sprinkle and Beth Stephens are concerned that their Ecosexual Manifesto and praxis address the consent of earthothers; see Stephens and Sprinkle "Ecosexuality," 2016.

3. I use "trans*" as Hayward and Weinstein (2015) suggest: trans* marks "the *with, through, of, in* and *across* that make life possible" (196); trans* is "'always already' relational" (198).

4. The commercial is now publicly available on YouTube, at http://www.youtube.com/watch?v=R3YHrf9fGrw

5. Mentioned here are terrorist attacks in the United States and Europe, only because my readers are most likely from these regions, but they are not the majority of those targeted. According to the Global Terrorism Index Report (2014), 82% of those killed in terrorist attacks died in just five countries: Iraq, Afghanistan, Pakistan, Nigeria, and Syria.

6. See Doug Vakoch, ed., *Transecology: Transgender Perspectives on the Environment* (2018) for an initial foray into this field.

7. Pulé's eight conceptual frameworks range from Progressive Left to Conservative Right, and include Socialist, Gay/Queer, Profeminist, Black (African), Mythopoetic, Men's Rights, Morally Conservative, and Evangelical; his seven liberatory ideals include Feminist Sociobiology, Deep Ecology, Social Ecology, Ecopsychology, Gaia Theory, Inclusionality Theory, and General Systems Theory.

8. Bruce Erickson (2010) introduces the Lesbian Rangers: "Shawna Dempsey and Lorri Millan founded the Lesbian National Parks and Services in 1997 as a way of inserting a lesbian presence into the natural landscape. In full uniform, the performance artists interact with the public, and point out potential hazards to the flourishing of lesbian flora and fauna in natural settings, including sexism and the naturalization of heterosexuality in human and nonhuman contexts" (Erickson 2010, 328 n. 3).

9. Social ecofeminist Chaia Heller (1999) describes five dimensions of the socio-erotic, humans' desire for sensuality, association, differentiation, development, and political opposition as part of an ecoanarcha-feminist eroto-politics.

10. Robert Bly's masculinist and anti-feminist work is purposely omitted as it does not advance ecofeminist politics.

11. See "Our Fight Too: An LGBTQ Response to Climate Change," at http://peoples-climate.org/lgbtq/2014/07/22/our-fight-too-an-lgbtq-response-to-climate-change-2/

12. For Stephens and Sprinkle's "Ecosex Manifesto," see http://sexecology.org/research-writing/ecosex-manifesto/ or their "Ecosexuality" (2016) essay.

Epilogue

If the world of nature dies, Wall Street dies too.

—Val Plumwood, Environmental Culture (2002, 236)

Critical ecofeminism's insights are not new. The ideas of our human interbeing with plants, animals, soil, rock, and sky that produces relational identities, not autonomous individualism; the animacy of all earthothers, reminding us that an animal's cries are not "the striking of a clock" (Descartes) but rather the pain of another living being who merits our care and responsiveness; the way that sharing (not privately accumulating) resources according to people's needs creates community harmony, because our economic, ecological, social, and material interdependence means that "we all do better when we all do better"[1]; the possibility that an ecoerotic celebrates our relations with life, or that trans* identities, as well relations of triads and multiples, occur across species—these and other ecojustice insights have appeared in diverse indigenous worldviews for millennia. Val Plumwood was careful not to appropriate indigenous knowledge, but rather to develop theory in conversation with specific indigenous thinkers, such as Native American writer Carol Lee Sanchez and Australian Aboriginal philosopher Bill Neidjie (Plumwood 2002). Almost every industrialized nation has an indigenous community as well as racially, ethnically, economically, or otherwise segregated communities whose experience and analyses hold ecological, economic, and sociopolitical significance for critical ecofeminists, for climate justice activists, and for interspecies justice activists alike.

During the month that this book neared completion, two "watershed" events took place. On November 8, 2016, a climate-denier candidate whose campaign was described by his opponent Hillary Clinton as "racist, sexist,

181

homophobic, xenophobic, Islamophobic, you name it," Donald Trump became the heir-apparent of the U.S. presidential election. In the days immediately following the election, while the United States and other nations grappled with the devastating effects Trump's presidency promised for public education, health care, human rights, animals and the environment, indigenous Dakota and Lakota Sioux were confronting oil colonialism on the Standing Rock reservation in North Dakota. On November 25, the U.S. Army Corps of Engineers sent a letter of eviction to the indigenous Water Protectors (not "protesters") at Standing Rock, camped near the Missouri River: their occupation of their own reservation lands, guaranteed in perpetuity by the 1868 Treaty of Fort Laramie, had become a final attempt to halt the Dakota Access Pipeline (Monet 2016; Sammon 2016). According to the eviction letter sent the day after Thanksgiving, a U.S. holiday of gratitude for the indigenous people who saved the lives of early European immigrants, the Standing Rock occupation would have to be removed by December 5, 2016.

From a critical ecofeminist perspective, the nexus of these events sows seeds for two different paths, leading either to our worst or best possible futures. How did these two events coarise? And what channels are most strategic for hope, resilience, and ecopolitical justice?

In April, May, and June of 2015, Hilary Clinton, Bernie Sanders, and Donald Trump (respectively) declared their candidacies for the U.S. presidency. A progressive Vermont Senator, Sanders quickly captured the optimism of voters concerned about climate justice, racial justice, gender, and sexual justice, as well as educational and economic justice. While Sanders' campaign was financed largely by small donations (averaging $25 per contribution) from a multitude of supporters, and addressed the structural economic causes of social ills in the U.S. and around the world, Trump's campaign was funded by the billionaire's own $100,000 in conjunction with millions of dollars from donors and political action committees, along with an estimated $55 million in free advertising through mainstream media coverage (Ingram 2016). Trump's campaign blamed Muslims, immigrants, and other marginalized groups for U.S. problems, promising that Trump would "Make America Great" again. While a Harvard study of campaign coverage confirms that Trump's campaign was highly profitable for the media (Ingram 2016), Sanders' message was rarely covered at all, and although Clinton received the most negative coverage for a candidate of either party, she still won the Democratic nomination, becoming the first woman candidate to earn major-party endorsement. But Clinton's mainstream ties to corporate funding and the Capital Hill "establishment," her support for fracking "under the right circumstances," and her leaked dismissal of climate activists in remarks that environmentalists should "get a life" did not sit well with young, progressive voters. Coupled with the quiet resentment of working-class rural whites, and the nonvoting decisions of the disillusioned, the presidential vote (amid concerns of ballot fraud and hacking) tilted toward Trump.

Connecting the presidential election to the oil pipeline through the leading candidates' climate insensitivity (Clinton) or climate denialism—particularly given Trump's between $500,000 and $1 million investment in Energy Transfer Partners, the parent company of the Dakota Access Pipeline (Milman 2016)—one finds the same economic and cultural system undermining rivers, native sovereignty, environmental justice, media democracy, and the climate.

On the ground in North Dakota, the U.S. Army Corps of Engineers went through the motions of seeking consent for the Dakota Access Pipeline from the Dakota and Lakota Sioux, who had repeatedly called for archaeological investigations, and whose calls were joined by the Osage nation and Iowa tribe, as their lands were also crossed by the proposed $3.8 billion, 1,172-mile pipeline that would carry 470,000 barrels of crude oil per day from North Dakota to southern Illinois (Estes 2016). In July 2016, the Army Corps of Engineers issued a fast-track permit to build the pipeline, ignoring objections that the route's crossing beneath the Missouri River, half a mile upstream from the Standing Rock reservation, would jeopardize drinking water, with an oil spill contaminating drinking water for millions downstream. By late August, activists had begun gathering at Standing Rock, with over 200 tribal nations from North America and Europe planting flags of solidarity around the camp. These activists recognized the rhetorical attempts to make indigenous people and sparsely-populated regions appear *remote*, and through their actions brought the ecosocial margins into the center of public attention.

But the bulldozers would not be stopped. On Labor Day weekend in September 2016, the Dakota Access bulldozers plowed a two-mile long trough through land that Standing Rock Sioux had publicly stated was archaeologically-significant and sacred tribal burial grounds. That weekend, the pipeline company hired private police who attacked native protesters with dogs and pepper-spray reminiscent of the 1950s civil rights battles—attacks that were recorded by progressive news reporters, such as Amy Goodman of DemocracyNow! A week later, North Dakota's Governor Jack Dalrymple called in the National Guard, and neighboring states sent their own police and Highway Patrol officers to quell the peaceful occupation so that the oil pipeline construction could proceed (thanks to overwhelming citizen demand, these police were soon recalled to their home states). On two separate occasions, the Department of Justice, the Department of the Army, and the Department of the Interior issued a joint statement, refusing to authorize construction in the Lake Oahe area near the peaceful occupation, and asking Energy Transfer Partners to cease construction—and the company refused. Their opportunity for economic profits would not be delayed by concerns for indigenous rights, ecological sustainability, or national governance.

On November 21, 2016, hundreds of water protectors were trapped on a bridge and attacked with water cannons, grenades, and bullets in freezing temperatures. Again, Amnesty International sent a fourth delegation out to Standing Rock to investigate events, and determined that U.S. law enforcement may have violated international law. Finally, this violent police encounter catapulted to national visibility the Veterans for Peace Call (2016) "for our fellow veterans to assemble as a peaceful, unarmed militia at the Standing Rock Indian Reservation on Dec 4–7 and defend the water protectors from assault and intimidation at the hands of the militarized police force and DAPL security." Where the prayers of nonviolent indigenous elders and water protectors had been unable to halt the militarized advance of oil colonialism, the nonviolent presence of U.S. military veterans sought to speak with more rhetorical force—and to do so on December 4, a day before the Thanksgiving letter's eviction deadline, December 5.

Organized by former Army officer Wes Clark Jr. and Marine Corps veteran Michael Wood Jr., the Standing Rock Veterans for Peace "Order for Operations" asserts, "First Americans have served in the United States Military, defending the soil of our homelands, at a greater percentage than any other group of Americans. There is no other people more deserving of veteran support" (Linehan 2016). As Wood explained, "we need to do the things that we actually said we're going to do when we took the oath to defend the Constitution from enemies foreign and domestic." In a section under "Friendly Forces," the VFP's Operations Order describes the mission's purpose:

> our intent is to honor the giants on whose shoulders we stand, such as Gandhi's salt protest or MLK's Selma protest. In the ultimate expression of alliance, we are there to put our bodies on the line, no matter the physical cost, in complete non-violence to provide a clear representation to all Americans of where evil resides. The Water Protectors are leading the way against this same evil which we must all face globally, saving ourselves and our children from the apocalyptic outcome of climate change. (2016)

Despite the humanist and masculinist-heroic overtones of combating "evil," the Order provides a clear recognition of indigenous rights and the intersections of oil colonialism and climate change. Although not a full articulation of feminist ecomasculinity, from the "deep time" perspective of stones, the Veterans for Peace support of Standing Rock water protectors recuperates the life-nourishing attributes ascribed to ecojustice standpoints and ecomasculinities: a commitment to nonviolence, an acknowledgment of elders and their exemplary leadership, solidarity with those oppressed, and a concern for the well-being of children and indigenous communities. These ecojustice values shift Western cultural (and militarized) masculinities from individualism and dominance to community and inhabitation.[2]

"This is what democracy looks like"—the chant begun at the people's resistance to the November 1999 World Trade Organization meeting in Seattle still resonates in the Veterans for Standing Rock commitment to "put our bodies on the line." And thus the dialectical interplay between progressive, intersectional on-the-ground ecojustice activism and the ideals of theory—whether Plumwood's critical ecofeminism, queer ecologies, anticolonial environmental justice, or indigenous cosmopolitics—continues. The numerous actions organized to challenge the corporate banks that are funding the Dakota Access Pipeline, the marches to local branches of the Army Corps of Engineers, the rallies and letters and editorials directed at elected officials, governmental bodies, and police, and the Women's March on Washington planned the day after the January 2017 presidential inauguration: embodied action offers an antidote for despair, for channeling grief and outrage at injustice, and for transforming the conditions of social-economic-environmental relations.

Art and narrative offer additional channels for hope, vision, and fresh perspectives capable of transforming political, economic and social relations. Feminist environmental artists and educators working through WEAD: Women Eco Artist Dialog, Artists 4 Climate at the Paris COP21 addressing climate change, Native artists, and many other eco-art activists offer alternative visions of community, energy, economics, and empowerment. Métis choreographer Rulan Tangen, director of Dancing Earth, has produced "Walking at the Edge of Water" to represent water's primal force and its life forms, and to address the many struggles around water issues (Hansen 2014). Navajo photographer Will Wilson has produced large-scale photographs depicting the distressed relationship between Native Americans and environments damaged by industrial intrusions. An artist-in-residence at the Institute of American Indian Arts in Santa Fe, New Mexico, Standing Rock-born artist Cannupa Hanska Luger has created mirrored shields for the water protectors, as has New York "social designer" Nikolas Bentel: both artists have produced simple videos for YouTube, instructing activists how to create mirrored signs and add simple slogans, such as "water is life" and "I stand with Standing Rock" (Angeleti 2016; Holmes 2016). In 2016, Minneapolis' BareBones Theater group performed its annual Halloween play on the theme of the Dakota Access Pipeline, complete with a papier-maché black snake and twenty bison, who later showed up to accompany protesters to the Army Corps of Engineers headquarters in St. Paul, much like the actual herd of bison appeared at Standing Rock in late October, as if the spirit guardians of native people were joining them in the battle to protect water (Cahill 2016).

Through these and other eco-art performances, climate justice enters the public imagination.

In *The Great Derangement: Climate Change and the Unthinkable* (2016), Amitav Ghosh explores Western culture's inability to comprehend the scale and violence of climate change. Cli-fi narratives project desolate, absurd and surrealistic futures that seldom equip readers for responding to climate change imperatives in the here and now, failing to provide accurate analyses of climate change causes or radically effective strategies for mitigating and adapting to climate change near-futures. History and politics alike fail us, Ghosh argues, providing gross simplifications or confining politics to the personal sphere. Climate change is a wicked problem because "the time horizon in which effective action can be taken is very narrow" (160), and requires swift community action from spiritual and political organizations. Ghosh's compelling argument urges writers and artists toward a transformation of literature and art, capable of breaking the spell of humanity's isolation and replacing us in kinship with other beings.

To find our way out of climate crises, we need to understand the root causes. As if in response to Ghosh's *Great Derangement*, Ta-Nehisi Coates' *Between the World and Me* (2015) narrates the story of collapse that intersects race, gender, sexuality, class, indigenous, and ecological colonialism in U.S. history, while Freya Mathews' *Ardea* (2016) narrates a story of recovery—from humanism's autonomous individualism, from Cartesian rationalism's mechanistic worldview, and from a capitalocentric approach to more ecological multispecies relations. Like other New Abolitionists in the United States (Hayes 2014), Coates narrates his autobiographical encounters with white supremacy, his re-education in blackness at Howard University, his fears for the safety of his own son, and concludes by comparing the political economy of slavery with the political economy of fossil fuel: these are the roots of climate change economics. An Australian environmental feminist philosopher, panpsychist, and literary scholar, Freya Mathews narrates a Faustian challenge set amid climate-changing forces of economic "development" predicated on ecosystem destruction, species loss, homophobia, and the loneliness of humanism. Here, in narratives from authors on three different continents, the collapse of individualism is identified as prerequisite to western culture's rediscovery of ecological kinship, powering and restorying our collective resistance and kincentric recovery in the Anthropocene.

Together, these tools of theory, art, narrative, and activism at all levels—cultural, ecological, economic, social, and transspecies—can change the story of climate injustice. Whether working on the level of international climate change agreements, developing local strategies for promoting community solar gardens or building multiracial alliances for racial, gender, and species justice, those who want to activate their own critical ecofeminist praxis have many avenues to choose.

NOTES

1. Minnesota's progressive and much-loved Senator (1991–2002) Paul Wellstone coined this phrase.

2. Veterans of many genders traveled to enact their solidarity with Standing Rock water protectors. Their presence and actions notwithstanding, the U.S. military as an institution has been shaped by the ideology of dominant masculinity and its associated features of heroism, conquest, violent "protection," warfare, nationalism, and identification of enemy/others.

Bibliography

A.I. Artificial Intelligence. Dir. Steven Spielberg. Dreamworks Studios, 2001.

Abadzis, Nick. *Laika.* New York: First Second, 2007.

"About AASHE." Association for the Advancement of Sustainability in Higher Education. http://www.aashe.org/about accessed on 1/5/2017.

Adams, Carol J. *The Pornography of Meat.* New York: Continuum Press, 2003.

———. "Woman-Battering and Harm to Animals," 55–84 in Carol J. Adams and Josephine Donovan, eds. *Animals & Women: Feminist Theoretical Explorations.* Durham, NC: Duke University Press, 1995.

———. "Comment on George's 'Should Feminists Be Vegetarians?'" *Signs* 21:1 (Autumn 1995), 221–225.

———. *The Sexual Politics of Meat: A Feminist-Vegetarian Critical Theory.* New York: Continuum, 1990.

———. ed. *Ecofeminism and the Sacred.* New York: Continuum, 1993.

———. and Josephine Donovan, eds. *Animals and Women: Theoretical Explorations.* Durham, NC: Duke University Press, 1995.

——— and Lori Gruen, eds. *Ecofeminism: Feminist Intersections with Other Animals and the Earth.* New York: Bloomsbury, 2014.

Adamson, Joni, and Cate Sandilands. "Vegetal Ecocriticism: The Question of 'The Plant.'" Preconference Seminar for the 2013 Conference of ASLE: Association for the Study of Literature and Environment. University of Kansas, Lawrence, KS. May 28, 2013. Accessed at http://asle.ku.edu/Preconference/adamson-sandilands.php on 1/5/2017.

Agarwal, Bina. "The Gender and Environment Debate: Lessons from India." *Feminist Studies* 18:1 (1992), 119–59.

Agostino, Ana, and Rosa Lizarde. "Gender and Climate Justice." *Development* 55:1 (2012), 90–95.

Aguilar, Lorena. "Women and Climate Change: Women as Agents of Change," *International Union for Conservation of Nature (IUCN)*, December 2007, available at http://cmsdata.iucn.org/downloads/climate_change_gender.pdf (accessed April 10, 2014).

———. Araujo, A. and Quesada-Aguilar, A. "Fact Sheet on Gender and Climate Change." International Union for Conservation of Nature (IUCN). Fact sheet presented at the UNFCCC COP 13, held in Bali in December 2007. Available at http://genderandenvironment.org/resource/gender-and-climate-change/ (accessed January 14, 2015).

Agyeman, Julian. "Toward a 'Just' Sustainability?" *Continuum: Journal of Media and Cultural Studies* 22:6 (December 2008), 751–756.

———. Robert D. Bullard and Bob Evans. "Exploring the Nexus: Bringing Together Sustainability, Environmental Justice and Equity." *Space & Polity* 6:1 (2002), 77–90.

——— and Bob Evans. "'Just Sustainability': The Emerging Discourse of Environmental Justice in Britain?" *The Geographical Journal* 170:2 (June 2004), 155–163.

Ahbez, Eden. *Eden's Island: The Music of an Enchanted Isle.* Hollywood, CA: Del-Fi Records LP, 1960.

Alaimo, Stacy. *Bodily Natures: Science, Environment, and the Material Self.* Bloomington, IN: Indiana University Press, 2010.

———. "Insurgent Vulnerability and the Carbon Footprint of Gender." *Kvinder, Kon & Forskning* [Women, Gender, & Research] 3 (2009), 22–35.

———. "Trans-corporeal feminisms and the ethical space of nature," 237–264 in *Material Feminisms*, edited by Stacy Alaimo and Susan Hekman. Bloomington, IN: Indiana University Press, 2008.

———. "'Skin Dreaming': The Bodily Transgressions of Fielding Burke, Octavia Butler, and Linda Hogan," 123–138 in *Ecofeminist Literary Criticism: Theory, Interpretation, Pedagogy*, edited by Greta Gaard and Patrick D. Murphy. Urbana, IL: University of Illinois Press, 1998.

———. and Hekman, Susan, eds. *Material Feminisms*. Bloomington, IN: Indiana University Press, 2008.

Alber, Gotelind, and Ulrike Roehr. "Climate Protection: What's Gender Got to Do with It?" *Women & Environments International Magazine* 70/71 (2006), 17–20.

Alexander, Michelle. *The New Jim Crow: Mass Incarceration in the Age of Colorblindness*. New York: The New Press, 2012.

Allen, John. *Biosphere II: The Human Experiment.* New York: Viking/Penguin Books, 1991.

———. and Mark Nelson. *Space Biospheres*. Oracle, AZ: Synergetic Press, 1989.

Allister, Mark, ed. *Eco-Man: New Perspectives on Masculinity and Nature.* Charlottesville: University of Virginia Press, 2004.

Alvares, Claude. *Another Revolution Fails: An Investigation into How and Why India's Operation Flood Project, Touted as the World's Largest Dairy Development Programme, Funded by EEC, Went off the Rails.* Delhi: Ajanta Publications, 1985.

An Inconvenient Truth: A Global Warning. Dir. Davis Guggenheim. Paramount Studios, 2006.

Anderlini-D'Onofrio, Serena. "Bisexuality, Gaia, Eros: Portals to the Arts of Loving." *Journal of Bisexuality* 11 (2011), 176–194.

Andersen, Kip, and Keegan Kuhn, dirs. *Cowspiracy: The Sustainability Secret.* Documentary film. AUM Media, 2014.

Anderson, Lorraine, ed. *Sisters of the Earth: Women's Prose and Poetry about Nature.* New York: Viking Books, 1991; 2003.

Anderson, William. *Green Man: The Archetype of our Oneness with the Earth.* London: HarperCollins Publishers, 1990.

Angeleti, Gabriella. "Artist creates mirrored shields for Standing Rock protesters." *The Art Newspaper*, November 21, 2016. Accessed at http://theartnewspaper.com/news/artist-creates-mirrored-shields-for-standing-rock-protesters/ on November 29, 2016.

"Annual Report: 2013." Association for the Advancement of Sustainability in Higher Education. http://www.aashe.org/files/aashe_annualreport2013.pdf accessed on 1/5/2017.

Anshelm, Jonas, and Martin Hultman. "A Green Fatwa? Climate Change as a Threat to the Masculinity of Industrial Modernity." *NORMA: International Journal for Masculinity Studies* 9:2 (2014), 84–96.

Appell, David. "Behind the Hockey Stick." *Scientific American*, February 21, 2005. Web.

Arie, India. "Better Way" and "Ghetto." *Testimony,* Vol. 2: Love and Politics. Universal Republic Records, 2009.

Armbruster, Karla. "'Buffalo Gals, Won't You Come Out Tonight': A Call for Boundary-Crossing in Ecofeminist Literary Criticism," 97–122 in *Ecofeminist Literary Criticism: Theory, Interpretation, Pedagogy*, edited by Gaard, G., and Murphy, P.D. (eds.) Urbana, IL: University of Illinois Press, 1998.

Artists 4 Climate: Paris 2015. Accessed at http://www.artists4climate.com/en/paris-2015/ on November 29, 2016.

Atwood, M. *MaddAddam,* New York: Random House, 2013.

———. *The Year of the Flood*, New York: Doubleday, 2009.

———. *Oryx and Crake*, New York: Random House, 2003.

Aviv, Rachel. "A Valuable Reputation: After Tyrone Hayes said a chemical was harmful, its maker pursued him." *The New Yorker.* February 10, 2014. Web. http://www.newyorker.com/reporting/2014/02/10/140210fa_fact_aviv?currentPage=all

Bacigalupi, Paolo. *The Windup Girl,* San Francisco: Night Shade Books, 2009.

Bali Principles of Climate Justice. August 29, 2002. Available at http://www.ejnet.org/ej/bali.pdf Accessed September 10, 2013.

Banerjee, Damayanti, and Michael Mayerfeld Bell. "Ecogender: Locating Gender in Environmental Social Science." *Society & Natural Resources* 20 (2007), 3–19.

Barad, Karen. *Meeting the Universe Halfway: Quantum Physics and the Entanglement of Matter and Meaning.* Chapel Hill, NC: Duke University Press, 2007.

Bauman, Whitney A. "The 'Eco-ontology' of Social/ist Ecofeminist Thought." *Environmental Ethics* 29:3 (2007), 279–98.

Belo, C. J., and R. M. Bruckmaier. "Suitability of low-dosage oxytocin treatment to induce milk ejection in dairy cows." *Journal of Dairy Science* 93 (2010), 63–69.

Bennett, Jane. *Vibrant Matter: A Political Ecology of Things.* Chapel Hill, NC: Duke University Press, 2010.

Benson, Melinda Harm and Craig, Robin Kundis. "The End of Sustainability." *Society and Natural Resources*, 27 (2014), 777–782.

Bergen, Lara. *A Story About Global Warming,* New York: Little Simon/Little Green Books, 2008.

Berry, Susan. "Planned Parenthood Black Community is blaming 'toxic masculinity' for the horrific shooting at a gay nightclub in Orlando, Florida." *Breitbart.com,* June 13, 2016. Accessed at http://www.breitbart.com/big-government/2016/06/13/black-planned-parenthood-orlando-toxic-masculinity/ on 8 August 2016.

Biehl, Janet. *Rethinking Ecofeminist Politics.* Boston: South End Press, 1991.

Birke, Lynda. "Exploring the Boundaries: Feminism, Animals, and Science," 32–54 in Carol J. Adams and Josephine Donovan, eds., *Animals and Women: Feminist Theoretical Explorations.* Durham, NC: Duke University Press, 1995.

Blair, Jeanine. "Media On Ice." *Feminist Media Studies.* Commentary and Criticism Section, Topic: "Beyond Global Warming," edited by Arthurs, J. and Zacharias, U. *Feminist Media Studies* 8:3 (2008), 318–323.

Blevins, Brooks. "Fireworking Down South." *Southern Cultures* 10:1 (Spring 2004), 25–49.

Bosch, Torie. "Climate change in The Hunger Games: How dystopian young-adult fiction is tackling the social consequences of global warming," *New America Foundation,* March 21, 2012, available at http://www.slate.com/articles/arts/future_tense/2012/03/the_hunger_games_birthmarked_delirium_ya_fiction_on_climate_change_.html (accessed April 10, 2014).

Bowens, Natasha. *The Color of Food / Brown Girl Farming.* Web. http://thecolorofood.org/ Accessed 2/8/2014.

Boyle, T. C. *A Friend of the Earth,* New York: Viking Books, 2000.

Brady, Judy. *1 in 3: Women with Cancer Confront an Epidemic.* San Francisco, CA: Cleis Press, 1991.

Braidotti, Rosi. *The Posthuman.* Oxford: Polity Press, 2013.

Brashares, Ann. *The Here and Now,* New York: Delacorte Press, 2014.

Bridges, Tristan, and Tara Leigh Tober. "Masculinity and Mass Shootings in the U.S." *The Society Pages.* July 23, 2015. Accessed at https://thesocietypages.org/feminist/2015/07/23/masculinity-and-mass-shootings/ on August 9, 2016.

Broad, William J. "As Biosphere is Sealed, Its Patron Reflects on Life." *The New York Times*: Science Section. September 24, 1991. Accessed on 11/5/2012.

Browne, Kath, and Catherine J. Nash, eds. *Queer Methods and Methodologies.* Surrey, England: Ashgate Publishing, 2010.

Brú Bistuer, Josepa, and Mercé Aguera Cabo. "A Gendered Politics of the Environment," 209–225 in Lynn A. Staeheli, Eleonore Kofman, and Linda J. Peake, eds. *Mapping Women, Making Politics: Feminist Perspectives on Political Geography.* New York: Routledge, 2004.

Bruckmaier, R.M. "Normal and disturbed milk ejection in dairy cows." *Domestic Animal Endocrinology* 29 (2005), 268–273.

Brundtland, Gro Harlem. *Our Common Future: The World Commission on Environment and Development.* New York: Oxford University Press, 1987.

Buege, Douglas. "Rethinking Again: A Defense of Ecofeminist Philosophy," 42–63 in Karen J. Warren, ed., *Ecological Feminism.* New York: Routledge, 1994.

Buell, Lawrence. *The Future of Environmental Criticism: Environmental Crisis and Literary Imagination*, Malden, MA: Blackwell Publishing, 2005.

Buerkle, C. Wesley. "Metrosexuality Can Stuff It: Beef Consumption as (Heteromasculine) Fortification." *Text and Performance Quarterly* 29:1 (January 2009) 77–93.

Bunch, Charlotte. *Passionate Politics: Feminist Theory in Action*. New York: St. Martin's Press, 1987.

Burgess, Colin, and Chris Dubbs. *Animals in Space: From Research Rockets to the Space Shuttle*. Chichester, UK: Springer/Praxis Books in Space Exploration, 2007.

Butler, Judith. *The Psychic Life of Power: Theories in Subjection*. Stanford, CA: Stanford University Press, 1997.

Butler, Octavia. *The Xenogenesis Trilogy,* New York: Time/Warner Books, 2000.

———. *Parable of the Sower,* New York: Warner Books, 1993.

Cahill, Tom. "Thousands of Wild Buffalo Appear Out of Nowhere at Standing Rock." *U.S. Uncut.com*, October 28, 2016. Accessed at http://usuncut.com/resistance/thousands-wild-buffalo-appear-nowhere-standing-rock/ on November 29, 2016.

Cain, Joey and Bradley Rose. "Who are the radical faeries?" Accessed at http://eniac.yak.net/shaggy/faerieinf.html on 10/16/2012.

Caldecott, Leonie, and Stephani Leland, eds. *Reclaim the Earth: Women Speak Out For Life On Earth*. London: The Women's Press, 1984.

Campbell, T. Colin, and Thomas M. Campbell. *The China Study*. BenBella Books, 2006.

CARE Canada. 2010. *Cyclone Nargis: Myanmar two years later*. Ottawa, CARE Canada. Web. http://care.ca/main/index.php?en&cyclonenargis

CARE. "Why Women and Girls?" Web. http://www.care.org/work/womens-empowerment/women Accessed 1/22/2014.

Carson, Rachel. *Silent Spring*, New York: Houghton Mifflin, 1962.

Cassidy, David with Kristin Davy. *One Small Step: The Story of the Space Chimps*. (57:00). Distributed by Victory Multimedia, Inglewood, CA. 1989.

Center for Progressive Reform. *International Environmental Justice and Climate Change*, 2008. Available at http://www.progressivereform.org/perspintlenvironJustice.cfm (accessed April 10, 2014).

Cheever, Holly. "Cow proves animals love, think, and act." Accessed at http://www.globalanimal.org/2012/04/13/cow-proves-animals-love-think-and-act/71867/ on July 25, 2012.

Chen, Mel Y., and Dana Luciano, eds. Special Issue: "Queer Inhumanisms: Has the Queer Ever Been Human?" *GLQ: A Journal of Lesbian and Gay Studies* 21:2–3 (2015).

Chemaly, Soraya. "In Orlando, As Usual, Domestic Violence was Ignored Red Flag." *Rolling Stone*, June 13, 2016. Accessed at http://www.rollingstone.com/politics/news/in-orlando-as-usual-domestic-violence-was-ignored-red-flag-20160613 on 11/24/2016.

Chester, Lynne. "Book Review: *Sustainable Capitalism: A Matter of Common Sense* and *Capitalism As If the World Matters*." *Journal of Radical Political Economics*, 43:3 (2011), 373–377.

Christ, Carol P. *Rebirth of the Goddess: Finding Meaning in Feminist Spirituality.* New York: Routledge, 1997.

———. and Judith Plaskow, eds. *Womanspirit Rising: A Feminist Reader in Religion.* New York: HarperCollins, 1979.

Clare, Cassandra. "Back From the Future: *The Here and Now* by Ann Brashares," *The New York Times*, April 4, 2014. Available at http://www.nytimes.com/2014/04/06/books/review/the-here-and-now-by-ann-brashares.html?_r=0 (accessed April 10, 2014).

Clorfene-Casten, Liane. *Breast Cancer: Poisons, Profits, and Prevention.* Monroe, ME: Common Courage Press, 1996.

Coates, Ta-Nehisi. *Between the World and Me.* New York: Spiegel & Grau, 2015.

Cohen, Jeffrey Jerome. "Stories of Stone," *postmedieval* 1:1/2 (2010), 56–63.

———. *Stone: An Ecology of the Inhuman.* Minneapolis: University of Minnesota Press, 2015.

Cohoon, Lorinda B. "Festive Citizenships: Independence Celebrations in New England Children's Periodicals and Series Books." *Children's Literature Association Quarterly* 31:2 (Summer 2006),132–53.

Colburn, Theo, Dianne Dumanoski, and John Peterson Myers. *Our Stolen Future.* New York: Penguin/Plume, 1996.

Collard, Andrée with Joyce Contrucci. *Rape of the Wild: Man's Violence against Animals and the Earth.* Bloomington, IN: Indiana University Press, 1989.

Collins, Patricia Hill. *Black Feminist Thought: Knowledge, Consciousness, and the Politics of Empowerment.* Boston: Unwin Hyman, 1990.

Collins, Suzanne. *Mockingjay.* Danbury, CT: Scholastic Press, 2010.

———. *Catching Fire.* Danbury, CT: Scholastic Press, 2009.

———. *The Hunger Games.* Danbury, CT: Scholastic Press, 2008.

"Conceivable Future: Climate Crisis = Reproductive Justice Crisis." At http://conceivablefuture.org/ accessed 11/24/2016.

Connell, R. W. *Masculinities.* Berkeley, CA: University of California Press, 1995.

Cooper, Marc. "'When He Hits You, It's a Compliment': Johnny Allen Rools, and Cult Members Buckle Under." *The Village Voice*, (April 2, 1991a), 26, 32.

———. "Take This Terrarium and Shove It." *The Village Voice,* (April 2, 1991b), 24–33.

———. "Profits of Doom: The Biosphere Project Finally Comes Out of the Closet—As a Theme Park." *The Village Voice*, (July 30, 1991c), 31–36.

———. "Faking it: The Biosphere Is a Model of the Earth After All—It's Suffering From Runaway Greenhouse Effect." *The Village Voice*, (November 12, 1991d), 19–21.

———."Biosphere 2, The Next Generation: All the Smithsonian's Horses and All the Smithsonian's Men Won't Put the Arizona Bubble Back Together Again." *The Village Voice,* May 5, 1992, 41–44.

Corea, Gena. *The Mother Machine: Reproductive Technologies from Artificial Insemination to Artificial Wombs.* New York: Harper & Row, 1985.

Crenshaw, Kimberlé. "Mapping the Margins: Intersectionality, Identity Politics, and Violence against Women of Color." *Stanford Law Review*, Vol. 43 (1991), 1241–1299.

———. "Demarginalizing the intersection of race and sex: a black feminist critique of antidiscrimination doctrine, feminist theory, and antiracist politics." *Chicago Legal Forum* (1989) 139–67.

Crichton, Michael. *State of Fear.* New York: HarperCollins, 2004.

Crittendon, Ann. *The Price of Motherhood: Why the Most Important Job in the World is Still the Least Valued.* New York: Henry Holt and Company, 2001.

Cuomo, Chris J. "Climate Change, Vulnerability, and Responsibility." *Hypatia* 26:4 (2011), 690–714.

———. *Feminism and Ecological Communities: An Ethic of Flourishing.* New York: Routledge, 1998.

———. "Unravelling the Problems in Ecofeminism," *Environmental Ethics* 15:4 (1992), 351–363.

Curtin, Deane. "Compassion and Being Human," 39–57 in Carol J. Adams and Lori Gruen, eds., *Ecofeminism: Feminist Intersections with Other Animals and the Earth.* New York: Bloomsbury, 2014.

———. "Toward an Ecological Ethic of Care." *Hypatia* 6:1 (Spring 1991), 60–74.

Danford, Natalie. "DaCapo Embraces Metrosexuality." *Publisher's Weekly* (January 29, 2004), 107.

Dankelman, Irene, ed. *Gender and Climate Change: An Introduction.* London: Earthscan, 2010.

Datar, Chhaya. *Ecofeminism Revisited: An Introduction to the Discourse.* India: Rawat Editions, 2011.

Davion, Vicki. "Is Ecofeminism Feminist?" in *Ecological Feminism*, ed. Karen J. Warren. New York: Routledge, 8–28.

Davis, Karen. "Thinking Like a Chicken: Farm Animals and the Feminine Connection," 192–212 in Carol J. Adams and Josephine Donovan, eds. *Animals & Women: Feminist Theoretical Connections.* Chapel Hill, North Carolina: Duke University Press, 1995.

De Cicco, Gabriela. "COP18: Between Losing Rights And Gender Balance." February 22, 2013. Association for Women's Rights in Development (AWID). Web. http://www.awid.org/News-Analysis/Friday-Files/COP18-Between-losing-rights-and-gender-balance

De Dios, Honorio Bartolomé. "Cayera," in J. Neil C. Garcia and Danton Remoto, eds., *Ladlad 3: An Anthology of Philippine Gay Writing.* Pasig: Anvil Publishing, 2007.

De La Cadena, Marisol. "Indigenous Cosmopolitics in the Andes: Conceptual Reflections beyond 'Politics.'" *Cultural Anthropology*, 25:2 (2010), 334–370.

Diamond, Irene. *Fertile Ground: Women, Earth, and the Limits of Control.* Boston: Beacon Press, 1994.

———. and Gloria Feman Orenstein, eds. *Reweaving the World: The Emergence of Ecofeminism.* San Francisco, CA: Sierra Club Books, 1990.

Di Chiro, Giovanna. "Sustaining Everyday Life: Bringing Together Environmental, Climate and Reproductive Justice." *DifferenTakes* 58. Spring (2009), 1–4.

Dickens, Peter. "The Cosmos as Capitalism's Outside." *Sociological Review* (2009), 66–82.

Dixon, Beth. "The Feminist Connection Between Women and Animals," *Environmental Ethics* 18:2 (1996), 181–194.

Dodd, Elizabeth. "The Mamas and the Papas: Goddess Worship, the Kogi Indians, and Ecofeminism." *NWSA Journal* 9:3 (1997), 77–88.

Donovan, Josephine. "Participatory Epistemology, Sympathy, and Animal Ethics," 75–90 in Carol J. Adams and Lori Gruen, eds., *Ecofeminism: Feminist Intersections with Other Animals and the Earth*. New York: Bloomsbury, 2014.

———. "Comment on George's 'Should Feminists Be Vegetarians?'" *Signs* 21:1 (Autumn 1995), 226–229.

———. "Animal Rights and Feminist Theory," *Signs* 15:2 (1990), 350–375.

——— and Carol J. Adams, eds. *The Feminist Care Tradition in Animal Ethics*. New York: Columbia University Press, 2007.

Duncan, Kirsty. "Feeling the Heat: Women's Health in a Changing Climate." *Canadian Women's Health Network*, Spring/Summer (2008). 4–7.

DuPuis, E. Melanie. *Nature's Perfect Food: How Milk Became America's Drink*. New York: New York University Press, 2002.

Duvernay, Ava. Dir. "13th." Netflix Documentary film. 2016.

Eaton, David. "Incorporating the Other: Val Plumwood's Integration of Ethical Frameworks," *Ethics & the Environment* 7:2 (2002), 153–180.

Ecesis.factor. "Just a Draft: Queer Justice, Climate Justice: Contemplating the Invisible Dimensions of Disaster in Hurricane Katrina." December 8, 2011. http://ecesisfactor.blogspot.com/2011/12/just-draft-queer-justice-climate.html

Edgerly, Mike. "MPCA Scientist Claims Harassment For Speaking Out About Chemicals." *Minnesota Public Radio Series "Toxic Traces,"* May 16, 2005. Accessed at http://news.minnesota.publicradio.org/features/2005/05/16_edgerlym_whistleblower/ on January 25, 2011.

"Editorial comment: Oxytocin, vasopressin and social behavior." *Hormones and Behavior* 61 (2012), 227–229.

Egeró, Bertil. "Population Policy: A Valid Answer to Climate Change? Old Arguments Aired Again Before COP15." *ACME: An International E-Journal for Critical Geographies*, 12:1 (2013), 88–101.

Ehrlich, Paul. *The Population Bomb*. Sierra Club/Ballantine Books, 1968.

Eisenstein, Zillah. *The Radical Future of Liberal Feminism*. New York: Longman, 1981.

Engelman, Robert. "Population, Climate Change, and Women's Lives." *WorldWatch Report #183*. Washington, D.C.: WorldWatch Institute, 2010.

"Environmental Justice." EPA: US Environmental Protection Agency. Accessed at https://www.epa.gov/environmentaljustice on 6/23/2016.

"Environmental Justice/Environmental Racism." *EJnet.org: Web Resources for Environmental Justice Activists*. Accessed at http://www.ejnet.org/ej on 6/23/2016.

Ergas, Christina, and Richard York. "Women's Status and Carbon Dioxide Emissions: A Quantitative Cross-National Analysis." *Social Science Research* 41 (2012), 965–976.

Erickson, Bruce. "'fucking close to water': Queering the Production of the Nation," 309–330 in Catriona Mortimer-Sandilands and Bruce Erickson, eds. *Queer*

Ecologies: Sex, Nature, Politics, Desire. Bloomington, IN: Indiana University Press, 2010.

Estes, Nick. "Fighting For Our Lives: NoDAPL in Historical Context." *The Red Nation.* September 18, 2016. Accessed at https://therednation.org/2016/09/18/fighting-for-our-lives-nodapl-in-context/ on November 28, 2016.

Estok, Simon. "Theorizing in a Space of Ambivalent Openness: Ecocriticism and Ecophobia." *Interdisciplinary Studies in Literature and Environment* 16.2 (2009), 203–225.

Evans, Arthur. *Witchcraft and the Gay Counterculture.* Boston: Fag Rag Books, 1978.

Feldman, Ruth. "Oxytocin and Social Affiliation in Humans." *Hormones and Behavior* 61 (2012), 380–391.

Ferrando, Francesca. "Posthumanism, Transhumanism, Antihumanism, Metahumanism and New Materialisms: Differences and Relations." *Existenz* 8:2 (2013), 26–32.

Fesher, Riham. "Proposed Sandpiper Oil Pipeline Moves Closer to Reality." *MPR News*, April 14, 2015. Accessed at http://www.mprnews.org/story/2015/04/14sandpiper-oil-pipeline on August 20, 2015.

Fiala, Nick. "How Meat Contributes to Global Warming." *Scientific American,* 4 February 2009. Available at http://www.scientificamerican.com/article.cfm?id=the-greenhouse-hamburger (accessed September 10, 2013).

Finn, John. "Masculinity and earth gods." *Certified Male: A Journal of Men's Issues* (Australia). Issue #9, 1998. Accessed at http://www.certifiedmale.com.au/issue9/earthgod.htm on 9/26/2012.

Flannery, Tim. *The Weather Makers.* New York: Atlantic Monthly Press, 2006.

Flocker, Michael. *The Metrosexual Guide to Style: A Handbook for the Modern Man.* Cambridge, MA: DaCapo Press, 2003.

Follman, Mark, Gavin Aronsen, and Deanna Pan. "A Guide to Mass Shootings in America." *Mother Jones,* July 18, 2016. Accessed at http://www.motherjones.com/politics/2012/07/mass-shootings-map?page=2 on 8/8/2016.

Food and Agricultural Association (FAO) of the United Nations. *State of Food Insecurity in the World 2013.* Web. http://www.fao.org/publications/sofi/en/

Fraiman, Susan. "Pussy Panic versus Liking Animals: Tracking Gender in Animal Studies." *Critical Inquiry* 39 (Autumn 2012), 89–115.

Fretwell, Holly. *The Sky's Not Falling! Why It's OK to Chill about Global Warming.* Los Angeles, CA: World Ahead Media, 2007.

Frick, Robert. "The Manly Man's Guide to Makeup and Metrosexuality." *Kiplinger's Report* (June 2004), 38.

Gaard, Greta. "Out of the Closets and into the Climate! Queer Feminist Climate Justice," in *Climate Futures: Re-Imagining Global Climate Justice.* Ed. Kum-Kum Bhavnani, John Foran, Priya A. Kurian, Debashish Munshi. Berkeley, CA: University of California Press, 2017.

———. "Speaking of Animal Bodies," *Hypatia* 27:3 (2012), 29–35.

———. "Ecofeminism Revisited: Rejecting Essentialism and Re-Placing Species in a Material Feminist Environmentalism," *Feminist Formations* 23:2 (Summer 2011), 26–53.

———. "New Directions for Ecofeminism: Toward a More Feminist Ecocriticism," *ISLE: Interdisciplinary Studies in Literature and the Environment* 17:4 (2010a), 1–23.

————. "Reproductive Technology, or Reproductive Justice? An Ecofeminist, Environmental Justice Perspective on the Rhetoric of Choice." *Ethics & the Environment* 15:2 (2010b), 103–129.

————. "Toward an Ecopedagogy of Children's Environmental Literature," *Green Theory & Praxis,* 4:2 (2009), 11–24.

————. *The Nature of Home: Taking Root in a Place.* Tucson, AZ: University of Arizona, 2007.

————. "Tools for a Cross-Cultural Feminist Ethics: Exploring Ethical Contexts and Contents in the Makah Whale Hunt," *Hypatia* 16:1 (2001), 1–26.

————. Rev. of Mary Zeiss Stange, *Woman the Hunter.* In *Environmental Ethics* 22:2 (2000), 203–206.

————. *Ecological Politics: Ecofeminists and the Greens.* Philadelphia: Temple University Press, 1998.

————. "Toward a Queer Ecofeminism." *Hypatia* 12:1 (Winter 1997), 114–137.

————. "Women, Animals, and an Ecofeminist Critique." *Environmental Ethics* 18:4 (1997), 440–43.

————. "Milking Mother Nature: An Eco-Feminist Critique of rBGH." *The Ecologist,* 24:6 (November/December 1994), 202–203.

————. "Review of Janet Biehl, *Rethinking Ecofeminist Politics.*" *Women and Environments,* 13:2 (1992), 20–21.

————. ed. *Ecofeminism: Women, Animals, Nature.* Temple University Press, 1993.

————. Simon Estok and Serpil Oppermann, eds. *International Perspectives in Feminist Ecocriticism.* New York: Routledge, 2013.

————. and Lori Gruen. "Comment on George's 'Should Feminists Be Vegetarians?'" *Signs: Journal of Women in Culture and Society* 21:1, (1995), 230–241.

Gagliano, Monica. "Seeing Green: The Re-discovery of Plants and Nature's Wisdom." *Societies* 3 (2013), 147–157.

Garbarino, James. "Protecting Children and Animals from Abuse: A Trans-Species Concept of Caring," 250–58 in Josephine Donovan and Carol J. Adams, eds. *The Feminist Care Tradition in Animal Ethics.* New York: Columbia University Press, 2007.

Garrard, Greg. *Ecocriticism,* New York: Routledge, 2004.

Gaye, Marvin. "Mercy, Mercy Me (The Ecology)." Motown Records, 1971.

Gender CC: Women for Climate Justice. Web. http://www.gendercc.net/Accessed 1/15/2004.`

George, Kathryn Paxton. "Should Feminists Be Vegetarians?" *Signs* 19:2 (1994), 405–434.

George, Susan. *How the Other Half Dies: The Real Reasons for Hunger.* NY: Penguin, 1976.

————. *Ill Fares the Land: Essays on Food, Hunger, and Power.* Washington, D.C.: Institute for Policy Studies, 1984.

Ghosh, Amitav. *The Great Derangement: Climate Change and the Unthinkable.* Chicago: University of Chicago Press, 2016.

Gibbs, Lois Marie. *Dying From Dioxin.* Boston, MA: South End Press, 1995.

Global Gender and Climate Alliance (GGCA). "GGCA at the UNFCCC COP19 Gender Workshop." Web. http://www.gender-climate.org/Events/COP19-Gender-Workshop.php

Global Terrorism Index Report 2014: Measuring and Understanding the Impact of Terrorism. New York, Sydney, and Oxford: Institute for Economics and Peace.

Goebel, Allison. "Women and Sustainability: What Kind of Theory Do We Need?" *Canadian Woman Studies* 23.1 (2004), 77–84.

Gosine, Andil. "Non-white Reproduction and Same-Sex Eroticism: Queer Acts against Nature," 149–172 in Catriona Mortimer-Sandilands and Bruce Erickson, eds. *Queer Ecologies: Sex, Nature, Politics, Desire.* Bloomington, IN: Indiana University Press, 2010.

———. "Pink Greens." *Alternatives*, 27:3 (July 2001). Accessed at http://www.alternativesjournal.ca/sustainable-living/pink-greens on June 20, 2015.

Gray, Elizabeth Dodson. *Green Paradise Lost.* Wellesley, MA: Roundtable Press, 1979.

Grewe-Volpp, Christa. "Keep Moving: Place and Gender in a Post-Apocalyptic Environment," 221–234 in Greta Gaard, Simon Estok, and Serpil Oppermann, eds. *International Perspectives in Feminist Ecocriticism.* New York: Routledge, 2013.

Griffin, Susan. *Woman and Nature: The Roaring Inside Her.* New York: Harper & Row, 1978.

Grubbs, Jenny, ed. Special Issue: "Inquiries and Intersections: Queer Theory and Anti-Speciesist Praxis." *Journal of Critical Animal Studies*, 10:3 (2012).

Gruen, Lori. "Facing Death and Practicing Grief," 127–141 in Carol J. Adams and Lori Gruen, eds., *Ecofeminism: Feminist Intersections with Other Animals and the Earth.* New York: Bloomsbury, 2014.

———. "Navigating Difference (again): Animal Ethics and Entangled Empathy," 213–234 in Gregory Smulewicz-Zucker, ed., *Strangers to Nature: Animal Lives & Human Ethics.* Lanham, MD: Lexington Books, 2012.

———. "Experimenting with Animals," 105–129 in *Ethics and Animals.* Cambridge University Press, 2011.

———. "Review of Janet Biehl, *Rethinking Ecofeminist Politics*," *Hypatia* 7:3 (1992), 216–20.

Halberstam, Judith. *Female Masculinity.* Durham, NC: Duke University Press, 1998.

Halcom, Chad. "Marathon Petroleum: Nearly 9 in 10 Property Owners Near Detroit Refinery Interested in Buyout Offer." *Crain's* April 11, 2012. Accessed at http://www.crainsdetroit.com/article/20120411/FREE/120419963/marathon-petroleum-nearly-9-in-10-property-owners-near-detroit-refinery-interested-in-buyout-offer 20 May 2012.

Hall, Jeremy K., Daneke, Gregory A., and Lenox, Michael J. "Sustainable development and entrepreneurship: Past contributions and future directions." *Journal of Business Venturing* 26 (2010), 439–448.

Hall, Matthew. *Plants as Persons: A Philosophical Botany.* Albany, NY: SUNY Press, 2011.

———. ed. Special Issue: "Plant Ethics." *PAN: Philosophy Activism Nature* 9, 2012.

Hansen, Teri. "Native Artists Use Works to Spark Environmental Awareness." *Indian Country Today Media Network*, February 24, 2014. Accessed at http://indiancountrytodaymedianetwork.com/print/2014/02/24/native-artists-use-works-spark-environmental-awareness-153731 on November 29, 2016.

Happy Feet. Dir. George Miller. Animal Logic/Warner Brothers, 2006.

Haraway, Donna. *The Companion Species Manifesto: Dogs, People, and Significant Otherness*. Chicago: Prickly Paradigm Press, 2003.

————. *Primate Visions: Gender, Race, and Nature in the World of Modern Science*. New York: Routledge, 1989.

Harding, Sandra. "Is There a Feminist Method?" *Feminism & Methodology*. Bloomington, IN: Indiana University Press, 1987. 1–14.

Harper, A. Breeze, ed. *Sistah Vegan: Black Female Vegans Speak on Food, Identity, Health, and Society*. New York: Lantern Books, 2010.

Harris Interactive. "LGBT Americans Think, Act, Vote More Green than Others." *Business Wire* (New York). October 26, 2009.

Hartmann, Betsy. 2009. "10 Reasons Why Population Control is not the Solution to Global Warming." *DifferenTakes* #57. Winter 2009.

————. *Reproductive Rights and Wrongs: The Global Politics of Population Control and Contraceptive Choice*. New York: Harper & Row, 1987.

———— and James Boyce. *Needless Hunger: Voices from a Bangladesh Village*. San Francisco: Institute for Food and Development Policy, 1979.

Harvey, Fiona. "Eat Less Meat To Avoid Dangerous Global Warming, Scientists Say." *The Guardian*, March 21, 2016. Accessed at https://www.theguardian.com/environment/2016/mar/21/eat-less-meat-vegetarianism-dangerous-global-warming on August 7, 2016.

Hawthorne, Susan. *Wild Politics*. North Melbourne, Australia: Spinifex Press, 2002.

Hayday, Matthew. "Fireworks, Folk-dancing, and Fostering a National Identity: The Politics of Canada Day." *The Canadian Historical Review*, 91:2 (June 2010), 287–314.

Hayes T.B., Collins A., Lee M., et al. "Hermaphroditic, demasculinized frogs after exposure to the herbicide atrazine at low ecologically relevant doses," *Proceedings of the National Academy of Sciences of the. U.S.A.* 99:8 (2002), 5476–80.

Hayes, Christopher. "The New Abolitionism." *The Nation*, April 22, 2014. Accessed at https://www.thenation.com/article/new-abolitionism/ on November 29, 2016.

Hayward, Eva, and Jami Weinstein. "Introduction: Tranimalities in the Age of Trans* Life." *TSQ: Transgender Studies Quarterly* 2:2 (May 2015), 195–206.

Hegland, Jean. *Into the Forest*. New York: Bantam/Calyx Books, 1996.

Heller, Chaia. *Ecology of Everyday Life: Rethinking the Desire for Nature*. Montreal: Black Rose Books, 1999.

Henry, Holly, and Amanda Taylor. "Re-thinking Apollo: Envisioning Environmentalism in Space." *Sociological Review*, 57:s1 (2009), 190–203.

Herreid, C. F. "Using Novels as Bases for Case Studies: Michael Crichton's State of Fear and Global Warming." *Journal of College Science Teaching* 34:7 (2005), 10–11.

Hillard, Richard. *Ham the Astrochimp*. Honesdale, PA: Boyds Mills Press, 2007.

Hogan, Katie. "Queer Green Apocalypse: Tony Kushner's Angels in America," 235–253 in Greta Gaard, Simon Estok, and Serpil Oppermann, eds. *International Perspectives in Feminist Ecocriticism.* New York: Routledge, 2013.

Holmes, Kevin. "Mirrored #NoDAPL Protest Signs Protect Protesters From Anti-Police." *The Creators Project*, November 19, 2016. Accessed at http://thecreatorsproject.vice.com/blog/mirrored-nodapl-protest-signs-protection-from-police on November 29, 2016.

Houle, Karen L. F. "Animal, Vegetable, Mineral: Ethics as Extension or Becoming? The Case of Becoming-Plant." *Journal for Critical Animal Studies* 9:1/2 (2011), 89–116.

Hubbard, Ruth. *The Politics of Women's Biology.* New Brunswick, NJ: Rutgers University Press, 1990.

Huggan, Graham. and Helen Tiffin. *Postcolonial Ecocriticism: Literature, Animals, Environment.* New York: Routledge, 2010.

Hultman, Martin. "The Making of a Modern Hero: A History of Ecomodern Masculinity, Fuel Cells and Arnold Schwarzenegger." *Environmental Humanities* 2 (2013), 79–99.

Ikerd, John E. *Sustainable Capitalism: A Matter of Common Sense. Bloomfield: Kumarian* Press, 2005.

Ingram, Mathew. "Here's Proof that The Media Helped Create Donald Trump." *Fortune*, June 14, 2016. Accessed at http://fortune.com/2016/06/14/media-trump/ on November 29, 2016.

Intergovernmental Panel on Climate Change (IPCC). N.d. "The IPCC Assessment Reports," available at http://www.ipcc.ch/ Accessed September 10, 2013.

International Lesbian, Gay, Bisexual and Trans and Intersex Association (ILGA). "Bolivian President: Eating Estrogen-Rich Chicken Makes You Gay," April 15, 2010. Available at http://ilga.org/ilga/en/article/moGN2RJ1vA Accessed September 10, 2013.

"The Island President." Dir. John Shenk. Afterimage Public Media, 2011.

Jaggar, Alison M., ed. *Living with Contradictions: Controversies in Feminist Social Ethics.* Boulder, CO: Westview Press, 1994.

jones, pattrice. "Fighting cocks: Ecofeminism versus sexualized violence." Pp. 45–56 in Lisa Kemmerer, ed. *Sister Species: Women, Animals, and Social Justice.* Chicago: University of Illinois Press, 2011.

———. "Stomping with the elephants: Feminist principles for radical solidarity." Pp. 319–333 in Steven Best and Anthony Nocella II, eds. *Igniting a Revolution: Voices in Defense of the Earth.* Oakland, CA: AK Press, 2006.

Johnson, L. "(Environmental) Rhetorics of Tempered Apocalypticism in 'An Inconvenient Truth.'" *Rhetoric Review,* 28:1 (2009), 29–46.

Kayne, Eric. "Defending Fenceline Communities from Oil Refinery Pollution." *EarthJustice* (2014). Case 2180, 3065. Accessed at http://earthjustice.org/our_work/cases/2014/defending-fenceline-communities-from-oil-refinery-pollution# on 6/23/2016.

Keller, Evelyn Fox. *Reflections on Gender and Science.* New Haven, CT: Yale University Press, 1985.

———— and Helen Longino, eds. *Feminism & Science*. New York: Oxford University Press, 1996.

Kelly, Petra. *Thinking Green! Essays on Environmentalism, Feminism, and Nonviolence*. Berkeley, CA: Parallax Press, 1994.

Kemmerer, Lisa, ed. *Sister Species: Women, Animals, and Social Justice*. Urbana, IL: University of Illinois Press, 2011.

Kemp, Martin. "A Dog's Life: Laika, the Doomed Stray, Has Achieved a Kind of Immortality." *Nature* 449 (October 4, 2007), 541.

Keon, Joseph. *Whitewash: The Disturbing Truth about Cow's Milk and Your Health*. Gabriola Island, BC: New Society Publishers, 2010.

Kheel, Marti. "Communicating Care: An Ecofeminist Perspective," *Media Development*, (February 2009), 45–50.

————. *Nature Ethics: An Ecofeminist Perspective*. Lanham, MD: Rowman & Littlefield, 2008.

————. "From Heroic to Holistic Ethics: The Ecofeminist Challenge," 243–271 in Greta Gaard, ed. *Ecofeminism: Women, Animals, Nature*. Philadelphia, PA: Temple University Press, 1993.

————. "From Healing Herbs to Deadly Drugs: Western Medicine's War Against the Natural World," 96–114 in Judith Plant, ed., *Healing the Wounds: The Promise of Ecofeminism*. Philadelphia, PA: New Society Press, 1989.

Kim, Claire Jean. "Slaying the Beast: Reflections on Race, Culture, and Species," *Kalfou* 1:1 (2010), 57–74.

————. "Multiculturalism goes Imperial—Immigrants, Animals, and the Suppression of Moral Dialogue," *Du Bois Review* 4:1 (2007), 233–49.

Kimmerer, Robin Wall. *Braiding Sweetgrass: Indigenous Wisdom, Scientific Knowledge, and the Teachings of Plants*. Minneapolis, MN: Milkweed Press, 2013.

King, Elaine A. "The Landscape in Art: Nature in the Crosshairs of an Age-Old Debate." *Artes Magazine* (November 16, 2010). Accessed at http://www.artesmagazine.com/2010/11/the-landscape-in-art-nature-in-the-crosshairs-of-an-age-old-debate/ on 1/25/2010.

Kingsolver, Barbara. *Flight Behavior*. New York: HarperCollins, 2012.

Klein, Naomi. *This Changes Everything: Capitalism vs. The Climate*. New York: Simon & Schuster, 2014.

————. "Geoengineering: Testing the Waters." *New York Times*, October 27, 2012. Accessed online at http://www.nytimes.com/2012/10/28/opinion/sunday/geoengineering-testing-the-waters.html on 11/6/2012.

Koehn, Peter H., and Juha I. Uitto. "Evaluating Sustainability Education: Lessons From International Development Experience." *Higher Education* 67 (2014), 621–635.

Kolbert, Elizabeth. *Field Notes From a Catastrophe: Man, Nature, and Climate Change*, New York: Bloomsbury, 2006.

Kolodny, Annette. *The Lay of the Land: Metaphor as Experience and History in American Life and Letters*. Chapel Hill, NC: University of North Carolina Press, 1975.

Kool Keith/Dr. Octagon. (2007) "Trees Are Dying," available at http://www.youtube.com/watch?v=XtsdtNdk5ao Accessed September 10, 2013.

Kostkowska, Justyna. *Ecocriticism and Women Writers: Environmentalist Poetics of Virginia Wolf, Jeanette Winterson, and Ali Smith.* London: Palgrave/Macmillan, 2013.

Kramarae, Cheris. *Women and Men Speaking.* London: Newbury House Publishers, 1981.

Kramb, Daniel. *From Here.* London: Lonely Coot, 2012.

Krupar, Shiloh R. "Transnatural Ethics: Revisiting the Nuclear Cleanup of Rocky Flats, CO, Through the Queer Ecology of Nuclia Waste." *Cultural Geographies* 19:3 (2012), 303–327.

Laclau, Ernesto, and Chantal Mouffe. *Hegemony and Socialist Strategy: Toward a Radical Democratic Politics.* London: Verso, 1985.

Lakoff, Robin. *Language and Women's Place.* New York: Harper & Row, 1975.

Lappé, Francis Moore, and Joseph Collins. *World Hunger: 12 Myths.* New York: Grove Press, 1998.

Leach, Melissa. "Earth Mother Myths and Other Ecofeminist Fables: How a Strategic Notion Rose and Fell." *Development and Change* 38:1 (2007), 67–85.

LeGuin, Ursula. *Buffalo Gals and Other Animal Presences/* Santa Barbara, CA: Capra Books, 1987.

———. *Always Coming Home.* New York: Harper & Row, 1985.

———. *The Word for World is Forest.* New York: Putnam Books, 1976.

LeMenager, Stephanie. *Living Oil: Petroleum Culture in the American Century.* New York: Oxford University Press, 2014.

———. and Stephanie Foote. "The Sustainable Humanities." *PMLA* 127:3 (2012), 572–278.

Leonard, Ann, ed. *Seeds: Supporting Women's Work in the Third World.* New York: The Feminist Press, 1989.

Leopold, Aldo. *A Sand County Almanac.* New York: Oxford University Press, 1949.

Levin, Sam. "At Standing Rock, women lead fight in face of arrest, Mace, and strip searches." *The Guardian*, November 4, 2016. https://www.theguardian.com/us-news/2016/nov/04/dakota-access-pipeline-protest-standing-rock-women-police-abuse?CMP=share_btn_fb Accessed 11/24/2016.

Lewis, Renee. "Life in Michigan's Dirtiest Zip Code." Al Jazerra America, March 3, 2014. Accessed at http://america.aljazeera.com/articles/2014/3/3/michigan-tar-sandsindustryaccusedofactingwithimpunity.html on April 1, 2014.

Li, Huey-li. "A Cross-Cultural Critique of Ecofeminism," 272–294 in Greta Gaard, ed. *Ecofeminism: Women, Animals, Nature.* Philadelphia, PA: Temple University Press, 1993.

Linehan, Adam. "'Where Evil Resides': Veterans Deploy to Standing Rock to Engage the Enemy—the U.S. Government," *Task & Purpose*, November 21, 2016. Accessed at http://taskandpurpose.com/where-evil-resides-veterans-deploy-to-standing-rock-to-engage-the-enemy-the-us-government/ on 11/28/2016.

Lloyd, Saci. *The Carbon Diaries 2015.* New York: Holiday House, 2010.

Luke, Timothy. "Reproducing Planet Earth? The Hubris of Biosphere 2." *The Ecologist*, 25:4 (1995), 157–61.

Lykke, Nina. "Non-Innocent Intersections of Feminism and Environmentalism." *Kvinder, Kon & Forskning* [Women, Gender & Research] 3–4 (2009). 36–44.

Lynch, Willie. "The Making of a Slave." 1712. Accessed at http://www.finalcall.com/artman/publish/Perspectives_1/Willie_Lynch_letter_The_Making_of_a_Slave.shtml on July 27, 2016.

Maathai, Wangari. *The Green Belt Movement*. New York: Lantern Books, 2004.

MacGregor, Sherilyn. "Only Resist: Feminist Ecological Citizenship and the Post-politics of Climate Change. *Hypatia: Special Issue on Climate Change* 29:3 (2014), 617–633.

———. "Gender and Climate Change: From Impacts to Discourses," *Journal of the Indian Ocean,* 6:2 (2010), 223–238.

———. "Feminist Perspectives on Sustainability," 467–492 in David V. J. Bell and Yuk-kuen Annie Cheung, eds., *Introduction to Sustainable Development: Encyclopedia of Life Support Systems*. Oxford, U.K.: United Nations Educational, Scientific and Cultural Organization, in partnership with EOLSS Publishers Co., Ltd., 2009.

———. *Beyond Mothering Earth: Ecological Citizenship and the Politics of Care*. Vancouver, BC: University of British Columbia Press, 2007.

Mader, Clemens, Geoffrey Scott, and Dzulkifli Abdul Razak. "Effective Change Management, Governance and Policy for Sustainability Transformation in Higher Education." *Sustainability Accounting, Management and Policy Journal* 4:3 (2013), 264–284.

Mann, Bonnie. *Women's Liberation and the Sublime: Feminism, Postmodernism, Environment*. New York: Oxford University Press, 2006.

Mann, Susan A. "Pioneers of U.S. Ecofeminism and Environmental Justice," *Feminist Formations* 23:2 (2011), 1–25.

Mannix, Andy, and Mike Mullen. "Milk Money: A Half-Million Cows Were Worth More Dead Than Alive, and Now We're All Paying the Price." *City Pages* 32:1626 (2012), 9–13.

Margolis, Cheryl. "Between Economic Justice and Sustainability," 277–292 in David N. Pellow and Robert J.Brulle, eds., *Power, Justice, and the Environment: A Critical Appraisal of the Environmental Justice Movement*. Cambridge, MA: MIT Press, 2005.

Masson, Jeffrey Moussaieff. *The Emperor's Embrace: Reflections on Animal Families and Fatherhood*. New York: Pocket Books, 1999.

Mathews, Freya. *Ardea: A philosophical novella*. Earth, Milky Way: Punctum Books, 2016.

Mayberry, Maralee, Banu Subramaniam, and Lisa H. Weasel, eds. *Feminist Science Studies: A New Generation*. New York: Routledge, 2001.

McCright, Aaron M. "The Effects of Gender on Climate Change Knowledge and Concern in the American Public." *Population and Environment* 32 (2010), 66–87.

McEwan, Ian. *Solar*. New York: Random House / Anchor Books, 2010.

McFarlane, Donovan A. and Agueda G. Ogazon. "The Challenges of Sustainability Education." *Journal of Multidisciplinary Research* 3:3 (Fall 2011), 81–107.

McGowan, Kat. "The Secret Language of Plants." *Quanta Magazine,* December 16, 2013. Web. http://www.simonsfoundation.org/quanta/20131216-the-secret-language-of-plants/

Miller, T. S. "Lives of the Monster Plants: The Revenge of the Vegetable in the Age of Animal Studies." *Journal of the Fantastic in the Arts*, 23:3 (2012), 460-479.

McGuinness, Mindy. *Not a Drop to Drink.* New York: Katherine Tegan Books, 2013.

McGuire, Cathleen, and Colleen McGuire. "Grass-Roots Ecofeminism: Activating Utopia,"186–203 in *Ecofeminist Literary Criticism: Theory, Interpretation, Pedagogy*, edited by Gaard, G., and Murphy, P.D. (eds.) Urbana, IL: University of Illinois Press, 1998.

McKibben, Bill. "A Movement for a New Planet." *TomDispatch.com*, August 18, 2013. Accessed at http://www.tomdispatch.com/blog/175737/ on July 29, 2016.

———. "Global Warming's Terrifying New Math." *Rolling Stone Magazine*. July 19, 2012. Accessed at http://www.rollingstone.com/politics/news/global-warmings-terrifying-new-math-20120719 on 12/3/2012.

———. "The Only Way to Have a Cow." *Orion Magazine*, March/April 2010. Accessed at https://orionmagazine.org/article/the-only-way-to-have-a-cow/ on August 10, 2016.

McMahon, Martha. "From the Ground Up: Ecofeminism and Ecological Economics." *Ecological Economics* 20 (1977), 163–173.

McMichael, Tony, Colin Butler and Haylee Weaver. "Position Paper on HIV and AIDS and Climate Change." Commissioned by the United Nations Environment Program (UNEP) and UNAIDS. February 2008.

McWilliams, James. "Agnostic Carnivores and Global Warming: Why Enviros Go After Coal and Not Cows." *Freakonomics.com*, 11/16/2011.

———. *Just Food: Where Locavores Get it Wrong and How We Can Eat Responsibly.* New York: Little, Brown & Co., 2010.

"Meditative Chimponaut." *Time* 78:23 (December 8, 1961), 52–53.

Meir, Christopher. "Fireworks." *Senses of Cinema* 27 (July/August 2003). Available at http://www.sensesofcinema.com/2003/cteq/fireworks/ accessed on 1/18/2012.

Mellor, Mary. *Feminism and Ecology.* Washington Square, NY: New York University Press, 1997.

———. *Breaking the Boundaries: Toward a Feminist Green Socialism.* London: Virago Press, 1992.

Merchant, Carolyn. *Earthcare: Women and the Environment.* New York: Routledge, 1995.

———. *The Death of Nature: Women, Ecology and the Scientific Revolution.* New York: Harper & Row, 1980.

Mies, Maria, and Vandana Shiva. *Ecofeminism.* London: Zed Books, 1993.

Miller, Laura. "Fresh Hell: What's behind the boom in dystopian fiction for young readers?" *The New Yorker*, 14 June 2010. Accessed at http://www.newyorker.com/arts/critics/atlarge/2010/06/14/100614crat_atlarge_miller?currentPage=all on April 10, 2014.

Milman, Oliver. "Dakota Access Pipeline Company and Donald Trump Have Close Financial Ties." *The Guardian.* Accessed at https://www.theguardian.com/

us-news/2016/oct/26/donald-trump-dakota-access-pipeline-investment-energy-transfer-partners on November 29, 2016.

Milne, Markus J., and Gray, Rob. "W(h)ither Ecology? The Triple Bottom Line, the Global Reporting Initiative, and Corporate Sustainability Reporting." *Journal of Business Ethics*, 118 (2013), 13–29.

Milsapps, Jan. *Screwed Pooch*. Booksurge Publishing, 2007.

Mohanty, Chandra Talpade. "'Under Western Eyes' Revisited: Feminist Solidarity through Anticapitalist Struggles." *Signs: Journal of Women in Culture and Society* 28:2 (2002), 499–535.

Molotch, Harvey. "Oil in Santa Barbara and Power in America." *Sociological Inquiry* 40 (Winter 1970), 131–144.

Monet, Jenni. "Army Corps Issues Eviction Notice to Standing Rock Sioux Tribe," *Indian Country Today*, 11/26/2016. Accessed at http://indiancountrytodaymedianetwork.com/2016/11/26/army-corps-issues-eviction-notice-standing-rock-sioux-tribe-166585 on 11/27/2016.

Moore, Niamh. *The Changing Nature of Eco/Feminism: Telling Stories from Clayoquot Sound*. Vancouver, BC: University of British Columbia Press, 2015.

Morris, Mark. "Sustainability: An Exercise in Futility." *International Journal of Business and Management,* 7:2 (2012), 36–44.

Mortimer-Sandilands, Catriona, and Bruce Erickson, eds. *Queer Ecologies: Sex, Nature, Politics, Desire*. Bloomington, IN: Indiana University Press, 2010.

Mosedale, Mike. "When Good Science is Bad Politics. *City Pages* (Minneapolis), March 29, 2006. Accessed at http://www.citypages.com/news/when-good-science-is-bad-politics-6690845 on January 25, 2011.

Murphy, Patrick D. "An Ecological Feminist Revisioning of the Masculinist Sublime." *Journal of the Canary Islands* 64 (Summer 2012), 79–94.

———. "Terraculturation, Political Dissolution, and Myriad Reorientations." *Tamkang Review,* 39:1 (2008), 3–18.

———. "Nature Nurturing Fathers in a World Beyond Our Control," 196–210 in Mark Allister, ed. *Eco-Man: New Perspectives on Masculinity and Nature*. Charlottesville: University of Virginia Press, 2004.

"The Nearest Thing." *Time* 77:7 (February 10, 1961), 60–61.

Nealon, Jeffrey. *Plant Theory: Biopower and Vegetable Life*. Palo Alto, CA: Stanford University Press, 2016.

Neimanis, Astrida. "Hydrofeminism: Or, On Becoming a Body of Water," *Undutiful Daughters: Mobilizing Future Concepts, Bodies and Subjectivities in Feminist Thought and Practice*, eds. Henriette Gunkel, Chrysanthi Nigianni and Fanny Soderback. New York: Palgrave Macmillan, 2012.

———. "feminist subjectivity, watered," *Feminist Review* 103 (2013), 23–41.

———. and Rachel Loewen Walker. (2014) "*Weathering*: Climate Change and the 'Thick Time' of Transcorporeality," *Hypatia* 29:3 (2014), 558–575.

Nelson, Julie A. "Feminism, Ecology and the Philosophy of Economics." *Ecological Economics* 20 (1997), 155–162.

Nhanenge, Jyette. *Ecofeminism: Towards Integrating the Concerns of Women, Poor People, and Nature into Development*. Lanham, MD: University Press of America, 2011.

Nikiforuk, Andrew. *Tar Sands: Dirty Oil and the Future of a Continent.* Vancouver, B.C.: Greystone Books, 2010.

Nixon, Rob. *Slow Violence and the Environmentalism of the Poor.* Boston: Harvard University Press, 2011.

Noske, Barbara. *Beyond Boundaries: Humans and Animals.* Montreal: Black Rose Books, 1997.

O'Brien, Caragh. *Birthmarked.* New York: Macmillan/Roaring Brook Press, 2010.

Ochoa, Maria. "Toxic Masculinity and the Orlando Pulse Shooting." In *Countering Hate with Knowledge, Fury, and Protest: Three Latina/o Studies Scholars Respond to Orlando Massacre,* at "Mujeres Talk," June 28, 2016. Accessed at https://library.osu.edu/blogs/mujerestalk/2016/06/28/countering-hate-with-love-latinao-scholars-respond-to-orlando-massacre/comment-page-1/#comment-115664 on August 8, 2016.

Oliver, Lauren. *Delirium.* New York: HarperCollins, 2011.

Olson, Lynne. *Freedom's daughters: The unsung heroines of the Civil Rights Movement from 1830 to 1970.* New York: Scribner, 2002.

Oreskes, Naomi, and Eric M. Conway. *The Collapse of Western Civilization: A View from the Future,* New York: Columbia University Press, 2014.

Orr, David W. *Ecological Literacy: Education and the Transition to a Postmodern World.* Albany, NY: State University of New York Press, 1992.

Oulette, Jennifer. "Space Dog Laika Finally Gets a Happy Ending." *DiscoveryNews,* July 12, 2011. Accessed at http://news.discovery.com/space/laike-the-russian-space-dog-finally-gets-a-happy-ending-110712.html on 11/17/2012.

Pearsall, Hamil, Joseph Pierce, and Robert Krueger. "Whither Rio + 20?: Demanding a Politics and Practice of Socially Just Sustainability." *Local Environment* 17:9 (October 2012), 935–941.

Peet, Bill. *The Wump World.* Boston, MA: Houghton Mifflin Co., 1970.

Pellow, David Naguib. "Toward a Critical Environmental Justice Studies: Black Lives Matter as an Environmental Justice Challenge." *Du Bois Review* 13:2 (2016), 221–236.

———. *Total Liberation: The Power and Promise of Animal Rights and the Radical Earth Movement.* Minneapolis, MN: University of Minnesota Press, 2014.

Pérez-Peña, Richard. "College Classes Use Arts to Brace for Climate Change." *The New York Times,* March 31, 2014, available at http://www.nytimes.com/2014/04/01/education/using-the-arts-to-teach-how-to-prepare-for-climate-crisis.html?_r=0 Accessed April 10, 2014.

Pew Commission on Industrial Farm Animal Production in America. *Putting Meat on the Table: Industrial Farm Animal Production in America.* April 29, 2008. Accessed online at http://www.pewtrusts.org/uploadedFiles/wwwpewtrustsorg/Reports/Industrial_Agriculture/PCIFAP_FINAL.pdf on January 7, 2013.

Pharr, Suzanne. *Homophobia: A Weapon of Sexism.* Little Rock, AR: The Women's Project, 1988.

Phillips, Mary, and Nick Rumens, eds., *Contemporary Perspectives on Ecofeminism.* New York: Routledge, 2015.

Piercy, Marge. *Woman on the Edge of Time,* New York: Fawcett Books, 1985.

Plant, Judith, ed. *Healing the Wounds: The Promise of Ecofeminism.* New Society Press, 1989.

Plumwood, Val. *The Eye of the Crocodile*, ed. Lorraine Shannon. Canberra: Australian National University E-Press, 2012.

———. "Nature in the Active Voice." *Australian Humanities Review*, #46. May 2009. Accessed at http://www.australianhumanitiesreview.org/archive/Issue-May-2009/plumwood.html on August 7, 2016.

———. "Shadow Places and the Politics of Dwelling." *Australian Humanities Review*, #44. March 2008. Accessed at http://www.australianhumanitiesreview.org/archive/Issue-March-2008/plumwood.html on 1/5/2017.

———. "The Cemetery Wars: Cemeteries, Biodiversity and the Sacred," 54–71 in Martin Mulligan and Yaso Nadarajah, eds., *Local-Global: Identity, Security and Community*, Vol. 3. (2007). Special issue: Exploring the legacy of Judith Wright.

———. "Journey to the Heart of Stone," 17–36 in Fiona Beckett and Terry Gifford, eds., *Culture, Creativity and Environment: New Environmentalist Criticism.* Amsterdam and New York: Rodopi, 2007.

———. "Decolonising Australian Gardens: Gardening and the Ethics of Place." *Australian Humanities Review* #36 (July 2005). Accessed at http://www.australianhumanitiesreview.org/archive/Issue-July-2005/09Plumwood.html on 1/5/2017.

———. "Gender, Eco-Feminism and the Environment." In Robert White, ed., *Controversies in Environmental Sociology*. Cambridge University Press, 2004. 43–60.

———. "Animals and Ecology: Towards a Better Integration." Working/Technical Paper, 2003. Australian National University Digital Collection. Accessed at http://hdl.handle.net/1885/41767 on 1/5/2017.

———. *Environmental Culture: The ecological crisis of reason.* New York: Routledge, 2002.

———. "Integrating Ethical Frameworks for Animals, Humans, and Nature: A Critical Feminist Eco-Socialist Analysis," *Ethics & Environment* 5:2 (2000), 285–322.

———. "Androcentrism and Anthropocentrism: Parallels and Politics." 327–355 in Karen J. Warren, ed. *Ecofeminism: Women, Culture, Nature.* Bloomington, IN: Indiana University Press, 1997.

———. "Being Prey," *Terra Nova* 1:3 (1996), 32–44.

———. *Feminism and the Mastery of Nature.* New York: Routledge, 1993.

———. "The Atavism of Flighty Females," *The Ecologist*, 22:1 (1992), 36.

———. "Nature, Self, and Gender." *Hypatia* 6:1 (Spring 1991), 3–27.

Pollan, Michael. "The Intelligent Plant: Scientists debate a new way of understanding flora." *The New Yorker.* December 23, 2013.

"PopOffsets: Smaller families, less carbon." At http://www.popoffsets.org/ accessed 1/6/2017.

Porritt, Jonathon. *Capitalism As If the World Matters.* London: Earthscan, 2007

Potts, Annie, ed. *Meat Culture.* Leiden, The Netherlands: Brill Human-Animal Studies, 2016.

———. and Donna Haraway. "Kiwi Chicken Advocate Talks with California Dog Companion." *Feminism and Psychology* 20:3 (2010), 318–336.

———. and Jovian Parry. "Vegan Sexuality: Challenging Heteronormative Masculinity through Meat-free Sex." *Feminism & Psychology* 20:1 (2010), 53–72.

Poynter, Jane. *The Human Experiment: Two Years and Twenty Minutes Inside Biosphere 2.* New York: Avalon Publishing Group/Thunder's Mouth Press, 2006.

"Principles of Environmental Justice." Available at http://www.ejnet.org/ej/principles. html Accessed 26 January 2013.

Probyn-Rapsey, Fiona, et al. "A Sustainable Campus: The Sydney Declaration on Interspecies Sustainability." *Animal Studies Journal* 5:1 (2016), 110–151.

Pskowski, Martha. "Is This the Future We Want? The Green Economy vs. Climate Justice." *DifferenTakes*: *A Publication of the Population and Development Program at Hampshire College* 79 (Spring 2013), 1–4.

Pulé, Paul M. "Caring for Society and Environment: Towards Ecological Masculinism." Paper Presented at the Villanova University Sustainability Conference, April 2009. Accessed at http://www.paulpule.com.au/Ecological_Masculinism.pdf on October 1, 2012.

———. "Ecology and Environmental Studies," 158–162 in Michael Flood, Judith Kegan Gardiner, Bob Pease, and Keith Pringle, eds. *Routledge International Encyclopedia of Men and Masculinities*. New York: Routledge, 2007.

Reed, T.V. "Environmental Justice Eco-Art." Accessed at http://culturalpolitics.net/ environmental_justice/art on November 29, 2016.

Ress, Mary Judith. *Ecofeminism in Latin America*. Maryknoll, NY: Orbis Books, 2006.

Resurrección, Bernadette P. "Persistent women and environment linkages in climate change and sustainable development agendas." *Women's Studies International Forum* 40 (2013), 33–43.

Revkin, Andrew. "Confronting the Anthropocene." *The New York Times*. May 11, 2011. Accessed at http://dotearth.blogs.nytimes.com/2011/05/11/confronting-the-anthropocene/ on 12/10/2012.

Ricciardelli, Rosemary, Kimberley A. Clow, and Philip White. "Investigating Hegemonic Masculinity: Portrayals of Masculinity in Men's Lifestyle Magazines." *Sex Roles* 63 (2010), 64–78.

Rich, Nathaniel. *Odds Against Tomorrow*, New York: Farrar Straus Giroux, 2013.

Robbins, John. *Diet for a New America*. Walpole, NH: Stillpoint Publishing, 1987.

Robinson, Kim Stanley. *Forty Signs of Rain,* New York: Bantam, 2004.

———. *Fifty Degrees Below,* New York: Bantam, 2005.

———. *Sixty Days and Counting,* New York: Bantam, 2007.

Rockwell, Anne. *Why Are the Ice Caps Melting? The Dangers of Global Warming*. New York: Collins, 2006.

Rohr, Ulrike. "Gendered carbon footprints—gendered mitigation policy." *Outreach,* November 27, 2012. DOHA 2012: UN Climate Change Conference (COP 18). Stakeholder Forum.

Rojas-Cheatham, Ann, Dana Ginn Parades, Shana Griffin, Aparna Shah, and Eveline Shen. "Looking Both Ways: Women's Lives at the Crossroads of Reproductive Justice and Climate Justice." Asian Communities for Reproductive Justice. *Momentum Series* 5.

Rose, Deborah Bird. "Val Plumwood's Philosophical Animism: Attentive Inter-Actions in the Sentient World," *Environmental Humanities* 3 (2013), 93–109.

Ruether, Rosemary Radford. *Gaia & God: An Ecofeminist Theology of Earth Healing*. San Francisco: HarperCollins, 1992.

————. *Sexism and God-Talk: Toward a Feminist Theology.* Boston: Beacon Press, 1983.

Salatino, Kevin. *Incendiary Art: The Representation of Fireworks in Early Modern Europe.* Getty Publications, 1997.

Salleh, Ariel. *Ecofeminism as Politics: nature, Marx and the postmodern.* London: Zed Books, 1997.

————. "Deeper Than Deep Ecology: The Eco-Feminist Connection." *Environmental Ethics* 6 (1984), 339–45.

Salmon, Enrique. "Sharing Breath: Some Links Between Land, Plants, and People." 196–210 in Alison H. Deming and Lauret E. Savoy, eds. *The Colors of Nature: Culture, Identity, and the Natural World.* Minneapolis, MN: Milkweed Publications, 2011.

Sammon, Alexander. "A History of Native Americans Protesting the Dakota Access Pipeline," *Mother Jones,* September 9, 2016. Accessed at http://www.motherjones.com/environment/2016/09/dakota-access-pipeline-protest-timeline-sioux-standing-rock-jill-stein on 11/27/2016.

Sandberg, L. Anders, and Tor Sandberg, eds. *Climate Change—Who's Carrying the Burden? The chilly climates of the global environmental dilemma.* Ottawa, Canada: Canadian Centre for Policy Alternatives, 2010.

Sandilands, Catriona. "Floral Sensations: Plant Biopolitics." 226–237 in Teena Gabrielson et al (ed.), *The Oxford Handbook of Environmental Political Theory.* New York: Oxford University Press, 2016.

————. "Desiring Nature, Queering Ethics: Adventures in Erotogenic Environments," *Environmental Ethics* 23 (2001), 169–188.

————. *The Good-Natured Feminist: Ecofeminism and the Quest for Democracy.* Minneapolis: University of Minnesota Press, 1999.

————. "Political Animals: The Paradox of Ecofeminist Politics." *The Trumpeter* 9 (Spring 1994), 90–96.

————. "Lavender's Green? Some thoughts on queer(y)ing environmental politics," *UnderCurrents* (May 1994), 20–24.

Santino, Jack. "Light up the Sky: Halloween Bonfires and Cultural Hegemony in Northern Ireland." *Western Folklore* 55.3 (Summer 1996), 213–231.

Sauvage, Julie. "'This Tableau Vivant . . . Might Be Better Termed a Nature Morte': Theatricality in Angera Carter's Fireworks." *Les Cahiers de la nouvelle/Journal of the short story in English,* 51(Autumn 2008), 123–136.

Sbicca, Joshua. "Eco-queer movement(s): Challenging heteronormative space through (re)imagining nature and food." *European Journal of Ecopsychology* 3 (2012): 33–52.

Schlosberg, David. *Defining Environmental Justice: Theories, Movements, and Nature.* New York: Oxford University Press, 2007.

Scholten, Bruce A. *India's White Revolution: Operation Flood, Food Aid and Development.* London: I. B. Tauris Publishers, 2010.

Schwalbe, Michael. "The Hazards of Manhood." *Yes!* 63 (Fall 2012), 42–44.

Seaberg, Kurt. *Artist's Statement.* Accessed at http://www.kurtseaberg.com/artist-statement/ on 10/22/2012.

Seager, Joni. "Noticing gender (or not) in disasters." *Geoforum* 37 (2006), 2–3.

Sen, Gita, and Caren Grown. *Development, Crises, and Alternative Visions: Third World Women's Perspectives.* New York: Monthly Review Press, 1987.

Seymour, Nicole. *Strange Natures: Futurity, Empathy, and the Queer Ecological Imagination.* Urbana, IL: University of Illinois Press, 2013.

Shiva, Vandana. *Staying Alive: Women, Ecology, and Survival in India.* London: Zed Books, 1989.

Siano, Brian. "The Skeptical Eye: Captain Future's Terrarium of Discipline." *The Humanist,* March/April 1992, 41–42.

Sietsema, Robert. "Five Reasons Why Manufacturing Human Breast Milk Cheese is Disgusting." *The Village Voice Blogs,* February 27, 2011. Accessed at http://blogs.villagevoice.com/forkintheroad/2011/02/five_reasons_wh.php on 6/29/2012.

Silliman, Jael, Marlene Fried, Loretta Ross, and Elena Guttierez. *Undivided Rights: Women of Color Organize for Reproductive Justice.* Boston: South End Press, 2004.

Singer, June. *Androgyny: Toward a New Theory of Sexuality.* New York: Anchor Books/Doubleday, 1977.

Singer, Peter. *Animal Liberation.* New York: Avon Books, 1975.

Siperstein, Stephen. "Climate Change Fiction: Radical Hope From an Emerging Genre." September 25, 2014. Available at http://eco-fiction.com/climate-change-fiction-radical-hope-from-an-emerging-genre/ Accessed January 14, 2015.

Slovic, Scott. "Science, Eloquence, and the Asymmetry of Trust: What's at Stake in Climate Change Fiction." *Green Theory & Praxis: The Journal of Ecopedagogy.* 4:1 (2008), 100–112.

Smith, Andrea. *Conquest: Sexual Violence and American Indian Genocide.* Cambridge, MA: South End Press, 2005.

———. "Ecofeminism Through an Anticolonial Framework," 21–37 in Karen J.Warren, ed., *Ecofeminism: Women, Culture, Nature.* Bloomington, IN: Indiana University Press, 1997.

Smith, Jordan Fisher. "Life Under the Bubble." *Discover Magazine*, October 20, 2010. Accessed online at http://discovermagazine.com/2010/oct/20-life-under-the-bubble on 11/5/2012.

Sohn, Emily. "Eco-friendly fireworks offer safer pyrotechnics." Discovery News (July 2, 2009). Accessed at http://dsc.discovery.com/news/2009/07/02/eco-friendly-fireworks.html on 1/20/2012.

Solotaroff, Paul. "In the Belly of the Beast: The Dirty Truth about Cheap Meat." *Rolling Stone*, December 10, 2013. Accessed at http://www.rollingstone.com/feature/belly-beast-meat-factory-farms-animal-activists on 11/24/2016.

Somera, Nina. "Que[e]r[y]ing the Climate Debates." *Women in Action* 2 (2009), 81–84.

Sontheimer, Sally. *Women and the Environment: A Reader; Crisis and Development in the Third World.* New York: Monthly Review Press, 1991.

Southern Decadence New Orleans. Available at http://www.southerndecadence.net/history.htm Accessed January 14, 2015.

Spender, Dale. *Man Made Language.* London: Routledge & Kegan Paul, 1980.

Spretnak, Charlene, ed. *The Politics of Women's Spirituality.* New York: Anchor Books/Doubleday, 1982.

Springmann, Marco, H. Charles J. Godfray Mike Rayner, and Peter Scarborough. "Analysis and valuation of the health and climate change cobenefits of dietary change." *Proceedings of the National Academy of Sciences of the USA*, 113:15, April 12, 2016. 4146–4151.

Sprinkle, Annie. "Beyond Bisexual," 103–107 in Loraine Hutching and Lani Kaahumanu, eds. *Bi Any Other Name: Bisexual People Speak Out*. New York: Alyson Publications, 1991.

Stange, Mary Zeiss. *Woman the Hunter*. Boston: Beacon Press, 1997.

Starhawk. *The Fifth Sacred Thing*, New York: Bantam Books, 1993.

Stein, Rachel. "Sex, Population, and Environmental Eugenics in Margaret Atwood's *Oryx and Crake* and *The Year of the Flood*," 184–202 in Greta Gaard, Simon Estok, and Serpil Oppermann, eds., *International Perspectives in Feminist Ecocriticism*. New York: Routledge, 2013.

Steinfeld, Henning, Pierre Gerber, Tom Wassenaar, Vincent Castel, Mauricio Rosales, and Cees De Haan. *Livestock's Long Shadow: Environmental Issues and Options*. Rome: Food and Agriculture Organization of the United Nations, 2006.

Steingraber, Sandra. *Having Faith: An Ecologist's Journey to Motherhood*. New York: Berkley Books, 2001.

———. *Living Downstream: An Ecologist Looks at Cancer and the Environment*. Boston, MA: Addison-Wesley Publishers, 1997.

Stengers, Isabelle. "Including Nonhumans in Political Theory: Opening Pandora's Box?" *Political Matter: Technoscience, Democracy, and Public Life*, (2010), 3–33.

———. "The Cosmopolitical Proposal." *Making Things Public: Atmospheres of Democracy*, (2005), 994–1004.

Stephens, Anne. *Ecofeminism and Systems Thinking*. New York: Routledge, 2015.

Stephens, Beth, and Annie Sprinkle. "Ecosexuality." In Renee C. Hoogland, ed. *Gender: Nature*. London: Macmillan Interdisciplinary Handbooks, 2016.

———. "Goodbye Gauley Mountain: An Ecosexual Love Story." Fecund Arts, 2013.

Stevenson, Bryan. *Just Mercy: A Story of Justice and Redemption*. New York: Spiegel & Grau, 2015.

Sturgeon, Noël. *Environmentalism in Popular Culture: Gender, Race, Sexuality, and the Politics of the Natural*. Tucson, AZ: University of Arizona Press, 2009.

———. *Ecofeminist Natures: Race, Gender, Feminist Theory and Political Action*. New York: Routledge, 1997.

Tara, Stephanie Lisa. *Snowy White World to Save*. Dallas, TX: Brown Books Publishing Group, 2007.

Tarter, Jim. "Some Live More Downstream than Others: Cancer, Gender, and Environmental Justice," 213–228 in Joni Adamson, Mei Mei Evans, and Rachel Stein, eds. *The Environmental Justice Reader: Politics, Poetics, and Pedagogy*. Tucson, AZ: University of Arizona Press, 2002.

Taylor, Dorceta. *The State of Diversity in Environmental Organizations: Mainstream NGOs, Government Agencies, Green 2.0*. July 20, 2014. Accessed at http://www.diversegreen.org/the-challenge/ on June 20, 2015.

Terry, Geraldine. "No climate justice without gender justice: An overview of the issues." *Gender & Development* 17:1 (2009), 5–18.

The Day After Tomorrow. Dir. Roland Emmerich. 20th Century Fox, 2004.

The Hunger Games. Dir. Gary Ross. Lions Gate Films, 2012.

Trans and Womyn's Action Camp 2012. Accessed at http://twac.wordpress.com/ on 12/31/2012.

Tuana, Nancy. "Gendering Climate Knowledge for Justice: Catalyzing a New Research Agenda," 17–31 in M. Alston and K. Whittenbury, eds. *Research, Action and Policy: Addressing the Gendered Impacts of Climate Change.* Dordrecht: Springer Media, 2013.

Tuhus-Dubrow, Rebecca. "Cli-Fi: Birth of a Genre." *Dissent* (Summer 2013). Available at http://www.dissentmagazine.org/article/cli-fi-birth-of-a-genre. Accessed 8 July 2014.

Twine, Richard. "Ecofeminism and Veganism: Revisiting the Question of Universalism." 191–207 in Carol J. Adams and Lori Gruen, eds. *Ecofeminism: Feminist Intersections with Other Animals and the Earth.* New York: Bloomsbury, 2014.

United Nations WomenWatch: Women, Gender Equality, and Climate Change. Web. http://www.un.org/womenwatch/feature/climate_change/ Accessed 1/11/2014.

University Leaders for a Sustainable Future. "The Talloires Declaration." 1990. Accessed at http://www.ulsf.org/programs_talloires.html on August 10, 2016.

Valentine, David. "Exit Strategy: Profit, Cosmology, and the Future of Humans in Space." *Anthropological Quarterly*, 85:4 (2012), 1045–1068.

Vallianatos, E. G., with McKay Jenkins. *Poison Spring: The Secret History of Pollution and the EPA.* New York: Bloomsbury Press, 2014.

Vieira, Patrícia, Monica Gagliano, and John Ryan, eds. *The Green Thread: Dialogues with the Vegetal World.* Lanham, MD: Lexington Books, 2016.

Veterans Stand for Standing Rock. *Operations Order*, December 4–7, 2016. Accessed at https://drive.google.com/file/d/0ByZLhosK39TpeDdyNWN4S0FTTlE/view on November 28, 2016.

Veysey, Lawrence. *The Communal Experience: Anarchist and Mystical Counter-Cultures in America.* New York: Harper & Row, 1973.

Vining, James. *First in Space.* Portland, OR: Oni Press, Inc., 2007.

Viveiros de Castro, Eduardo. "Exchanging Perspectives: The Transformation of Objects into Subjects in Amerindian Ontologies." *Common Knowledge* 10:3 (2004), 463–484.

Von Keyserlingk, Marina A.G., and Daniel M.Weary. "Maternal behavior in cattle." *Hormones and Behavior* 52 (2007), 106–113.

Walker, Alice. *In Search of Our Mothers' Gardens.* New York: Harcourt, Inc., 1983.

Wall-E. Dir. Andrew Stanton. Pixar Animation Studios, 2008.

Waring, Marilyn. *If Women Counted: A New Feminist Economics.* New York: HarperCollins, 1988.

Warren, Karen. *Ecofeminist Philosophy: A Western Perspective on What It Is and Why It Matters.* Lanham, MD: Rowman & Littlefield Publishers, Inc., 2000.

———. "The Power and The Promise of Ecological Feminism." *Environmental Ethics* 12 (1990), 125–144.

———. ed. *Ecofeminism: Women, Culture, Nature.* Bloomington, IN: Indiana University Press, 1997.

————. ed. *Ecological Feminism.* New York: Routledge, 1994.

Waterworld. Dir. Kevin Reynolds. Universal Pictures, 1995.

Watts, Richard. "Towards an ecocritical postcolonialism: Val Plumwoood's *Environmental Culture* in dialogue with Patrick Chamoiseau," *Journal of Postcolonial Writing,* 44:3 (2008), 251–261.

WEAD: Women Eco Artists Dialog. Accessed at http://weadartists.org/ on November 29, 2016.

Weary, Daniel M., Jennifer Jasper, and Maria J. Hotzel. "Understanding Weaning Distress." *Applied Animal Behaviour Science* 110 (2007), 24-41.

Weisberg, Zipporah. "The Broken Promises of Monsters: Haraway, Animals, and the Humanist Legacy," Journal of Critical Animal Studies 7:2 (2009), 21–61.

Werrett, Simon. *Fireworks: Pyrotechnic Arts & Sciences in European History.* Chicago: University of Chicago Press, 2010.

White, Lynn, Jr. "The Historical Roots of Our Ecologic Crisis." *Science,* 155:3767 (1967), 1203–1207.

Wiley, Andrea S. *Re-Imagining Milk.* New York: Routledge, 2011.

Williams, Terry Tempest. *Desert Quartet: An Erotic Landscape.* New York: Pantheon Books, 1995.

Winterson, Jeanette. *Written on the Body.* New York: Vintage Books, 1992.

Wolfe, Cary. *What is Posthumanism?* Minneapolis: University of Minnesota Press, 2010.

————. "Human, All too Human: 'Animal Studies' and the Humanities." *PMLA: Publications of the Modern Language Association* 124.2 (2009) 564–75.

Women's Environment and Development Organization (WEDO). 2012. *Celebrating Momentum and Milestones.* Web.

Women's Environmental Network (U.K.). *Women's Manifesto on Climate Change,* May 15, 2007. Available at http://www.wen.org.uk/climatechange/resources/manifesto.pdf Accessed April 10, 2014.

Woodbury, Mary Sands. "Exploring Eco- and Climate Fiction." Available at http://eco-fiction.com/climate-fiction/ Accessed January 14, 2014.

Woolf, Virginia. *Orlando.* New York: Harcourt Brace Jovanovich, 1928; 1956.

World Food Programme (WFP). "Ten Facts about Women and Hunger." March 5, 2013. Web. http://www.wfp.org/stories/10-facts-about-women-and-hunger Accessed 2/8/2014.

World Health Organization. "Gender, Climate Change, and Health." Web. http://www.un.org/womenwatch/feature/climate_change/ Accessed 1/11/2014.

Wright, Laura Wright. *The Vegan Studies Project: Food, Animals, and Gender in the Age of Terror.* Athens, GA: University of Georgia Press, 2015.

————. *Wilderness into Civilized Shapes: Reading the Postcolonial Environment.* Athens, GA: University of Georgia Press, 2010.

Wuerker, Matt. "Michigan 2016." *Politico.com,* January 5, 2016. Accessed at http://www.politico.com/gallery/2016/01/matt-wuerker-political-cartoons-january-2016–002168?slide=5 on July 25, 2016.

Young, Iris Marion. *Justice and the Politics of Difference.* Princeton, NJ: Princeton University Press, 1990.

Ziser, Michael and Julie Sze. "Climate Change, Environmental Aesthetics, and Global Environmental Justice Cultural Studies," *Discourse* 29:2, 3 (2007), 384–410.

Index

350.Org, 109, 110, 111, 139, 173

Adams, Carol, x, xiv, xv, xvi, xxiv,
        xxvii, 28, 32, 129, 164;
    absent referent, 34
Adamson, Joni, 27, 28, 33, 39,
        43n1, 44n5
Agarwal, Bina, xv
Agyeman, Julian, 8, 24n5
ahbez, eden, 172
Alaimo, Stacy, 31, 33, 154, 158.
    See also transcorporeality
Alvares, Claude, 57–59
Anger, Kevin, 81, 82
animacy, xxi, xxviin9, 28, 39, 181
animal studies, 27, 28, 33–36, 40–41,
        43n3, 51, 93, 163;
    critical, xxi, xxiv, 17, 31–35, 41,
        44n4, 51, 65, 101, 151;
    feminist, xxiii, 33–35, 41, 92, 99,
        129, 130, 144;
    human-animal, 51, 87;
    vegan studies, xvi, 61
anthropocene, 91, 109, 147, 186;
    ecofeminism, xxvin2;
    feminism, x
Anzaldúa, Gloria, 33
apocalypse, 75, 88, 159n2

Bali Principles of Climate Justice,
        133–34, 137, 138, 153, 159n10
Barad, Karen, viii, xxii
Bennett, Jane, xxii, 21
Biehl, Janet, xv
bisexuality, 162, 171
Bloom, Dan, ix, 145, 159n2
breast cancer and environment, 129,
        141n7, 154
breastfeeding, 50, 53, 66;
    breastmilk, 52;
    economic value, 53–55;
    family leave, 54;
    gift economy, 53, 55, 57;
    lactivist, 55;
    Mother's Milk Project, 49, 51;
    race and class differences, 54–55,
        67n1
Buddhism, 41–43
Bunch, Charlotte, 122
Burke, Edmund, 75, 107

cannibalism, 40, 45n10, 55, 56
capitalism, 5, 21, 41, 106, 139, 146,
        152, 159n6, 168, 174
carnism, 28
Carson, Rachel, 33, 129, 154, 158n1
Carter, Angela, 80, 82

Cartesian dualisms, 20, 87;
    culture/nature, xxv, 15, 44, 52, 73,
        80, 170;
    erotophobia, 136;
    essentialist, 22, 34, 44;
    gender, 170;
    human-nature, xxiv, 108, 111;
    identity, 30;
    rationalism, xix, 186
cli-fi narratives, 143–53;
    adaptation and survival, 146;
    children's and young adult
        narratives, 148–50;
    documentary and film, 150, 155, 158;
    empathy, 143–44, 150;
    environmental humanities, 144, 151;
    environmental sciences, 144, 153;
    feminist environmental writing, 144;
    fiction, 145;
    heroism, 145, 150–51;
    nonfiction, 147–48;
    truncated narrative, 152–53.
See also climate justice narratives
climate change, viii, 4, 9, 18, 23,
        45n9, 88, 91, 105, 107–09,
        173, 184;
    critical ecofeminism, 157–58;
    divestment, 139, 141n11;
    feminist political ecology, 121;
    gender, 117–18;
    GLBTQ impacts, 124–26, 135–36,
        137, 179n11;
    Hurricane Katrina, 124–25;
    immigration, 126, 128;
    impacts on women, 123–24, 140n4;
    industrial revolution, 122;
    militarism, 128;
    overconsumption, 117, 126, 127,
        131, 139;
    population, 126–27, 133;
    root causes, 157;
    space exploration, 109–11;
    women's activism, 118–21, 125;
    Women's Environment and
        Development Organization
        (WEDO), 119–20, 126, 134, 139

climate justice:
    Copenhagen Climate Conference, 117;
    distributive ethics 117;
    gender and sexual justice, 126;
    Lady Justice, 117–18;
    masculinism, 118;
    queer, 125, 137, 141n5;
    techno-science, 118, 121
climate justice narratives, 154–57;
    "Cayera," 155–56;
    India Arie, 156;
    The Island President, 155;
    Kool Keith, 157;
    "Mercy, Mercy Me," 156
Coates, Ta-Nehisi, 186
Cohen, Jeffrey Jerome, xxi
Colburn, Theo, 129, 154
colonialism, xx, 16–17, 25n21, 54, 70,
        81, 87, 99, 125, 138, 144, 145,
        174, 186;
    oil, xxi, 14–15, 182, 184;
    "progress," xxvii, 50, 52, 57, 58
contextual moral veganism/
        vegetarianism, xxiv, xxviii;
    arguments, 37–38, 40–43;
    critical ecofeminism, 38
Cook, Katsi, 49, 51, 52
Cowspiracy, 14, 37, 141n8
critical ecofeminism, *xxii–xx*vi:
    anti-colonial, 14, 15–17, 25n21,
        44n5, 185;
    indigenous self-determination,
        182–85;
    relational self, 21;
    transspecies communication, xxi
Curtin, Deane, xxiv, xxv, xxviiin12,
        37, 38
dairy:
    Bovine Growth Hormone (rBGH),
        49, 56;
    cows, 49, 56–59, 62–66;
    farmers, 49, 56, 59;
    free rider, 57;
    industry, 49, 51, 56, 61, 62, 68n7;
    racism, 51, 60–62.
See also milk

Derrida, Jacques, 27
Descartes, René, xxv, 41, 181
Donovan, Josephine, xv, xvii, 34

Earth Day, 13
earthothers, viii, xiii, xix, xx, xxii,
        xxv, xxvi, xxviin1, xxviin7,
        14, 15, 21, 22, 25n21, 176,
        178n2, 181
ecocriticism, ix, xvi, 144, 151, 166;
   environmental justice, 152;
   feminist, xxiii, xxvii, 144,
      151, 152, 177;
   postcolonial, xvi, xxii, 17;
   vegan/vegetarian, 28;
   vegetal, 27, 39, 41, 43n1, 43n3
ecofeminism, 3, 4, 38, 161, 166, 170, 177;
   animals, xv, 98;
   anti-colonialism, 15–17, 44n5;
   biodiversity, xv;
   branches of theory, xv, 126, 160n13;
   critical, xii, xiv, xv, xvi, xxi,
      xxii–xxvi, 38, 39, 41, 43, 144,
      152, 157–58, 181, 185;
   critique of western
      environmentalisms, xv;
   ecological citizenship, xv, 21, 140;
   environmental justice, xv;
   essentialism, xv, 126, 158;
   globalization, xv;
   intersectional analyses, xvi;
   postcolonial, xvi, 17, 34, 52;
   posthumanism, xv;
   queer, xxviiin11, 39, 174;
   species justice, xxiv–xv;
   sustainability, 18–21;
   theory stages, xiii–xvi.
   See also listening
ecogender, 109, 161, 162, 167, 169,
      170–71, 178
ecological literacy, 23n1
ecomasculinity, x, 167, 172–74, 184;
   feminist, 168–69;
   queer, 170;
   Veterans for Peace, 184–85
economic man, 19–20

ecophobia, 28, 88, 136
ecoqueer, 132, 154, 175, 176
ecosexuality, x, 176, 177, 178n2, 179
empire, 15, 44.
   See also fireworks
environmental humanities, xxiii, 7, 20,
      33, 101, 144, 151
environmental justice, viii, xvi, xxii,
      9–15, 16–19, 23n1, 24n14,
      159, 183, 185;
   climate, 133, 138, 153, 155;
   coal mining, 175–76;
   feminist, xv, xvi, xxii, 129, 154;
   fireworks, 84;
   industrial animal food production,
      130, 132;
   milk, 52;
   oil, 9–11;
   ovarian cancer, 173;
   plant studies, 43;
   Plumwood, Val, xi, xxvii, 21;
   Principles, 8, 11, 133, 153;
   "sacrifice zones," 12;
   space exploration, 86, 101, 114n16;
   speciesism and racism, 37;
   sustainability, 3, 4, 7–8, 22, 24n5.
   See also environmental racism;
      reproductive justice; water
environmental racism, viii, 11, 12, 51,
      85.
   See also racism
erotophobia, 88, 136, 171, 174
Estok, Simon, ix, xxvii.
   See also ecophobia

farmworkers, vii–viii
feminist theory:
   animal studies, xxiii, 33–35, 41, 92,
      99, 129, 130, 144;
   communication studies, xvii;
   economics, 20, 53–55, 137;
   intersectionality, 9, 19, 139, 151;
   material, xxii, 33, 128, 144, 157,
      158;
   methodology, xvii, 19, 34–35
fireworks, 69–89;

animals, 69, 73, 85;
art, 71–74;
Canada, 77–78;
Catholicism, 71, 73, 75;
China, 70, 85;
class distinctions, 76, 79–80;
colonialism, 81;
elitism, 70, 73;
empire, 73, 82, 88;
environmental racism, 84–85;
European history, 70–77;
India, 85;
macchina, 71, 75, 76;
materials, 83–86;
militarism, 72, 74;
monarchy, 70–75;
nationalism, 77, 78, 81, 88n2;
nature, 72, 73, 74, 75, 80;
Northern Ireland, 77;
propaganda, 71, 74;
slow violence, 85, 87, 88;
spectacle, 69, 70–77, 87;
sublime, 75–77;
United States, 69, 77, 78–80;
violence against women, 80–81
food justice, viii, 101, 132, 138;
interspecies, 132;
queer, 132
food studies, 49–51, 61

Garrard, Greg, xxviii n. 10
gender justice, 20, 52, 126, 133,
137, 153.
See also ecogender; ecomasculinity
George, Shanti, 58, 60
Ghosh, Amitav, 159n1, 186
Gibbs, Lois, 129, 154
The Giving Tree, 66
Goodall, Jane, 29, 97, 100
Gruen, Lori, x, xvi, xxvii, 38, 97, 130

Haraway, Donna, 51, 93, 94,
112n2, 130;
humananimals, 108;
naturecultures, 33
Hayes, Tyrone, 154

Heller, Chaia, xv, 179n9
humanism, xxviin6, 30, 35, 39, 41,
44n4, 45n9, 99, 112n2, 128,
139, 153, 186
hunger, ix, 14, 36, 37, 131, 132, 133;
The Hunger Games, 149

interspecies justice, x, xvi, xxiv,
xxviiin12, 20, 43, 52, 128,
132, 133, 160n13, 181

jones, pattrice, ix, 65, 67n4

Kelly, Petra, xiv
Kemmerer, Lisa, xvi, 66
Kheel, Marti, x, xiv, xv, 19, 38;
ethics, 99, 109, 152;
masculinism, 93, 109, 111, 114n18,
139, 164, 171;
sacrifice, 15, 112n3;
truncated narrative, 87, 101, 133,
152, 153
Kim, Claire Jean, xxiv
Kimmerer, Robin Wall, xviii
kincentric, viii
King, Ynestra, xiv
Klein, Naomi, 108, 159n6
Kohn, Eduardo, 33, 39

LaDuke, Winona, xiii, 33, 49, 109
Leopold, Aldo, 20;
land ethic, 3, 23n1, 111, 114n18
listening, 110, 119;
antiracist multispecies relations,
xvi–xxii, 18, 41;
feminist methodology, 18, 22, 41,
98, 99
Livestock's Long Shadow, 14, 37, 131,
153, 181
Lorde, Audre, 41

MacGregor, Sherilyn, 21–22,
121, 125, 127, 128,
140n4, 145;
ecological citizenship, xv, 140
Marder, Michael, 32, 33, 41, 42

masculinity, dominant hetero-, xxiii,
    78, 93, 98, 108, 109, 135, 139,
    161, 163, 177, 187n2;
  ecological, 163, 165–74, 184;
  Green Man, 165;
  heroic, 92, 94, 101, 107, 110, 145,
    150, 184;
  lethal, 164–65;
  meat-eating, 164;
  patriarchal religions, 165;
  Radical Faeries, 44n4, 169–70;
  Violence, 164–65.
  See also ecogender; ecomasculinity;
    ecosexuality; space exploration
master model. *See* Plumwood, Val
Mathews, Freya, xxiii, 186
McKibben, Bill, 109–10, 138
meat, ix, xxv, 18, 28, 33–43, 67n4,
    93, 104;
  carnism, 43n2;
  climate change, 110, 111;
  climate justice, 131;
  ecological costs, 132, 150;
  human health, 131;
  hunger, 132.
  See also masculinity
Mellor, Mary, xiv
milk, 49–68;
  animal science studies, 62–66;
  colonialism, 57–60;
  commodification, 50;
  cows, 50;
  Europe, 50;
  feminist environmental science, 51;
  India, 49, 50, 57–60, 67n2;
  narrative, 50;
  nursing, 66, 68n7;
  'the perfect food,' 50;
  racism, 51, 60–62.
  See also dairy
Murphy, Patrick D., 76, 107, 151, 174

nature, xvi, 16, 21, 72, 93, 109, 140,
    162, 165, 166, 170, 173, 174,
    177;
  animal, 57, 73, 74;

capitalism, 5, 25n23, 134, 157;
  domination, 4, 50, 71, 74, 75, 76, 80,
    87, 91, 108, 156, 165, 174;
  gardening, 42;
  gender essentialisms, 121;
  mother, ix, 57, 66, 119, 151, 170,
    177, 181;
  queer, 132, 136, 170;
  techno-science, 91, 93
Neimanis, Astrida, xiii, 154
new materialisms, 45n9;
  agency, 28;
  plants and animals, 29, 38
Noske, Barbara, xv

Oppermann, Serpil, ix, xxviin5, xxviin9
Our Common Future, 4, 119, 121

Pellow, David, 17
plant studies, xxii, 27;
  genealogies, 32–33;
  posthumanist methodologies, 30;
  queer methodologies, 30–31
Plato, xxv, 41
Plumwood, Val, xxiii–xxvi;
  agency, xiii;
  critical bioregionalism, 20;
  ecological animalism, xv, xxiv,
    xxviiin12;
  ecological citizenship, 21;
  gardening, 42;
  indigenous thinkers, 181;
  Master Model, xxii–xxiv, 16, 19,
    25n21, 35, 43n2, 52, 81, 87, 128,
    136, 140, 152;
  materialist spirituality of place, 20;
  multispecies communication, xvi,
    xix, xxi–xxii;
  ontological vegetarianism, xxiv,
    xxvin3;
  place principle of environmental
    justice, 21;
  postcolonial theory, xvi;
  posthumanism, xxviin6;
  remoteness, 16, 18.
  See also earthothers

Pollan, Michael, 27, 28, 30
posthumanism, xv, xvi, xxiii, xxiv,
　　　　xxvin2, xxviin6, 3, 4, 27, 30,
　　　　32, 33, 35, 41, 44, 51, 111,
　　　　118, 128, 134, 137, 138, 139,
　　　　140, 141n11, 144, 160n13, 176
Potts, Annie, viii

queer, 3, 17, 33, 124, 132, 136,
　　　　140, 140n1, 141n5,
　　　　141n11, 142n12, 147, 167,
　　　　170, 176, 177;
　　activism, 44n4;
　　animals, 31, 52;
　　botanical, 31;
　　cli-fi, 154–55;
　　climate justice, 125, 134, 137, 175;
　　ecofeminism, xxviiin11, 39, 174;
　　ecology, xxviiin11, 31, 132, 135,
　　　　166, 169, 175, 185;
　　ecomasculinities, 170;
　　environmental justice, 176–77;
　　feminist, 35, 118, 125, 137,
　　　　162, 177;
　　food justice, 132, 133, 138;
　　gender, 135, 139;
　　methodology, 30;
　　multiversalism, 31;
　　plant studies, 39;
　　relational ontologies, 175–76;
　　studies, 30, 44n4, 171;
　　vegetarians, 166

racism:
　　environmental, viii, 11, 12, 51, 85;
　　institutional, xii, 25n26;
　　intersectionalities with sexism,
　　　　speciesism, classism, 15, 34, 37,
　　　　43n1, 52, 54, 67n1, 99, 111, 167;
　　milk, 61–62;
　　space exploration, 94, 106;
　　unlearning, xx
reproductive justice, 128, 130, 132,
　　　　133, 138
Rigby, Kate, xxiii, 45n8

rock, xix, xvi, xxii, xxvi, xxviin9, 15,
　　　　39, 42, 92, 162, 175, 178;
　　deep time, 184
Rose, Deborah Bird, xvii, xix, xxiii

sacrifice:
　　animal, 15–16, 80, 112n3;
　　slaves, 25n22;
　　space exploration, 93, 95, 99;
　　zones, 12, 15
Sandilands, Catriona ("Cate"):
　　ecological citizenship and
　　　　democracy, xv, 140, 25n27;
　　plant studies, 27, 28, 31, 32, 33,
　　　　43n1;
　　queer ecology, xxviiin11, 132, 135,
　　　　166, 169, 174, 175, 176
Seaberg, Kurt, viii, 172, 173
Shiva, Vandana, xiv, 52, 67n2
Singer, June, 166
Singer, Peter, 27, 33, 36, 99
space exploration:
　　animals, 92–100, 104, 105, 113n7;
　　anti-ecological, 93, 99, 101,
　　　　103–5, 109;
　　Biosphere II, 100–105;
　　climate change, 109–11;
　　colonialism, 93, 103, 107;
　　earthanimalities, 108, 109;
　　elitism, 91, 100, 104, 105–06;
　　feminist philosophy of science, 93,
　　　　98–99;
　　geoengineering, 107–08;
　　human identity, 108, 111;
　　Laika, 95–96, 97, 98;
　　masculinism, 92, 93, 98, 101,
　　　　107, 108;
　　meat-eating, 106, 110–11;
　　Military-industrial-space complex,
　　　　106;
　　nationalism, 94, 96, 99;
　　NewSpace corporations, 106–08;
　　racism, 94, 106;
　　sustainability, 101, 104.
　　See also techno-science

speciesism, ix, xv, 15, 28, 34, 37, 43n1, 44n6, 67n4, 97, 99, 125, 144, 153
Standing Rock, 182–85, 187n2
Steingraber, Sandra, 49, 51, 129, 154, 173
Stephens, Beth, and Annie Sprinkle, x, 142n12, 175–78, 178n2, 179n12;
    "Goodbye Gauley Mountain," 175–77
stone. *See* rock
Sturgeon, Noël, xv; 16, 17, 140n1, 160n13
sustainability, xvi, xvii, xxii, xxiii, xxvi, 3–25, 87;
    Association for the Advancement of Sustainability in Higher Education (AASHE), 5–7, 9;
    capitalism, 5;
    definitions, 4–9, 19–20;
    as development, 4;
    ecological, 76, 91, 104, 169, 174, 183;
    economic, 19, 86;
    environmental science and technology, 7;
    interspecies, 17;
    Just Ecofeminist Sustainability, 22–23;
    sustainability studies, xxii, xxiii, 5, 7, 8, 14;
    triple-bottom-line, 3–5;
    white, male, middle-class, 3

techno-science, 75;
    cli-fi narratives, 147, 149, 156, 158;
    climate justice, 118, 125, 130, 137, 140;
    space exploration, 91–93, 98, 103, 105, 106

transcorporeality, xxiii, xxvin1, 19, 22, 35, 38, 39, 41, 128, 140, 175.
    See also Alaimo, Stacy
transgender, xvii, 44n4, 124, 125, 162, 167, 170, 178n3;
    transecology, 166
Twine, Richard, xxiv, 44n7

veganism, xv, xvi, xxv, 27, 28, 31, 34, 36–43, 44n7, 51, 61, 110, 141n8, 152, 160n13, 166, 167;
    arguments for, 36;
    carnism, 43n2;
    contextual moral, 38, 41, 43
Viveiros de Castro, Eduardo, 33, 39–41

Waring, Marilyn, 137
Warren, Karen, xiv, 34, 42, 44n6, 168
water, xx, xxi, 7, 10:
    animacy, 29;
    animal food production, 37, 57, 131, 132;
    bodies, xiii, xix, 175;
    climate change, 146, 149, 150, 156;
    ecosexuality, 162, 176;
    environmental justice, 11–12;
    fireworks, 69, 72, 73, 83, 84;
    Flint, MI, 11, 24n12;
    indigenous, xiii, xviii, 9, 182–85, 187n2;
    industrial pollution, 5, 9, 12, 13, 18, 25n16, 152, 154;
    infrastructure, 86;
    powdered milk, 49, 57, 64, 67n1;
    space exploration, 96, 102;
    women, 121, 123, 137
Wolfe, Cary, xxviiin6, 30
Wright, Laura, ix, xvi

About the Author

Greta Gaard is professor of English and coordinator of the Sustainability Faculty Fellows at the University of Wisconsin-River Falls. Her critical eco-feminism emerges from the intersections of feminism, environmental justice, queer studies, and critical animal studies, exploring a wide range of issues, from interspecies justice, material perspectives on fireworks and space exploration, to postcolonial ecofeminism, and the eco-politics of climate change. Her creative nonfiction eco-memoir, *The Nature of Home*, has been translated into Chinese and Portuguese.